Y0-BCJ-169

About Island Press

Island Press, a nonprofit organization, publishes, markets, and distributes the most advanced thinking on the conservation of natural resources—books about soil, land, water, forests, wildlife, and hazardous and toxic wastes. These books are practical tools used by public officials, business and industry leaders, natural resource managers, and concerned citizens working to solve both local and global resource problems.

Founded in 1978, Island Press reorganized in 1984 to meet the increasing demand for substantive books on all resourcc-rclated issues. Island Press publishes and distributes under its own imprint and offers these services to other nonprofit organizations.

Support for Island Press is provided by Geraldine R. Dodge Foundation, The Energy Foundation, The Charles Engelhard Foundation, The Ford Foundation, Glen Eagles Foundation, The George Gund Foundation, William and Flora Hewlett Foundation, The John D. And Catherine T. MacArthur Foundatino, The Andrew W. Mellon Foundation, The Joyce Mertz-Gilmore Foundation, The New-Land Foundation, The J. N. Pew, Jr. Charitable Trust, Alida Rockefeller, The Rockefeller Brothers Fund, The Rockefeller Foundation, The Florence and John Schumann Foundation, The Tides Foundation, and individual donors.

About World Wildlife Fund

World Wildlife Fund is the largest private U.S. organization working worldwide to conserve nature. WWF works to preserve the diversity and abundance of life on Earth and the health of ecological systems by protecting natural areas and wildlife populations, promoting sustainable use of natural resources, and promoting more efficient resource and energy use and the maximum reduction of pollution. WWF is affiliated with the international WWF network, which has national organizations, associates, or representatives in nearly forty countries. In the United States, WWF has more than one million members.

GETTING AT THE SOURCE

GETTING AT THE SOURCE

Strategies for Reducing Municipal Solid Waste

WORLD WILDLIFE FUND

WWF®

The Final Report of the Strategies for Source Reduction Steering Committee

ISLAND PRESS

Washington, D.C. ❑ Covelo, California

This book is printed on recycled paper. The text paper is Recycle 100. It contains no dyes other than those in the wastepaper used to make the paper. The postconsumer-paper content of Recycled 100 can vary from 10 to 50 percent, depending on the individual batch of paper used for printing, with the remainder being mill-waste stock. The cover paper is Cross Pointe Genesis. It includes a minimum 15 percent postconsumer stock, with the remainder being mill-waste stock. During the recycling process, it has been bleached using sodium hypochlorite.

Agri-Tek vegetable-oil-based inks have been used in printing this book. These inks contain 35 to 40 percent corn, soy, and other vegetable oils and approximately 8 percent petroleum (compared to approximately 20 percent petroleum content for standard offset inks). Vegetable oils have a reduced Volatile Organic Compound (VOC) content, and are renewable resources, unlike petroleum oils. As with all offset inks, some chemicals on EPA's Toxics Release Inventory were added to the inks as drying agents.

This information is provided to help identify some of the considerations involved in evaluating materials used in publications. The combination of papers and inks used in this particular book reflects trade-offs among environmental attributes of materials, printing quality, price, and availability within specified time periods.

Library of Congress Cataloging-in Publication Data

Getting at the source : strategies for reducing municipal solid waste / World Wildlife Fund
p. cm.
Includes bibliographical references
ISBN 1-55963-163-5 (acid-free paper).—ISBN 1-55963-162-7 (pbk : acid-free paper)
1. Source reduction (Waste management) I. World Wildlife Fund (U.S.)
TD793.95.G47 1992
363.72'8—dc20 92-6124
CIP

Printed on recycled, acid-free paper

Manufactured in the United States of America

10 9 8 7 6 5 4 3 2 1

Contents

Preface

Source reduction policy, particularly as it applies to municipal solid waste, is evolving rapidly. Not long ago, public concern over our nation's dwindling landfill capacity helped catapult source reduction to the top of the waste management hierarchy. But additional concerns have emerged to give a new imperative for source reduction today, including the need to conserve our resources, to reduce the toxicity of wastes, and to improve business competitiveness. This combination of motivations has led to remarkable partnerships of people—in both public and private sectors—who are working together to achieve progress in source reduction.

In the summer of 1989, World Wildlife Fund & The Conservation Foundation (WWF) launched its Strategies for Source Reduction Project to explore opportunities for reducing municipal waste at the source, with particular emphasis on the product component of the waste stream. The objectives of the project were to analyze systematically opportunities for changes in product design and use that would accomplish waste reduction and to provide a forum for building consensus on policies to encourage reduction of municipal solid waste.

This project has attempted to draw into the process as many perspectives about municipal waste source reduction as possible. It has been guided by a 19-member Steering Committee composed of experts from government, industry, and public interest groups who represent a wide range of perspectives on municipal waste source reduction. While each and every recommendation may not be ideal from each member's perspective, this report reflects the Steering Committee's consensus that the entire set of recommendations, taken as a whole, provides a sound springboard for future debate and action.

The Steering Committee and WWF would like to express their sincere appreciation to the U.S. Environmental Protection Agency, the John D. and Catherine T. MacArthur Foundation, the William and Flora Hewlett Foundation, and the individual and corporate supporters of WWF for their generous support of this effort. Particular thanks go to Paul Kaldjian, EPA project director. Finally, we would like personally to thank Fran Irwin and Leah Haygood for their contributions as co-directors during the first half of this project, Bradley Rymph for his editorial assistance, and Joy Patterson for providing staff support throughout the project.

Gail Bingham
Christine Ervin

Project Co-directors

Steering Committee Membership

Although the Steering Committee was constituted to reflect the variety of views among interest groups concerned about municipal waste source reduction, the members of the Steering Committee listed below participated as individuals, not as official representatives of their organizations or institutions.

Frank Aronhalt, Director of Environmental Affairs for Plastics, E. I. du Pont de Nemours & Company

Bill Brown, Director of Environmental Affairs, Waste Management, Inc.

Linda Bruemmer, Deputy Director, Minnesota Office of Waste Management

Armin Clobes, Section Manager, Research and Development, S.C. Johnson Wax

Richard Denison, Senior Scientist, Environmental Defense Fund

Randall Franke, County Commissioner, Marion County, Oregon

Diana Gale, Director, Seattle Solid Waste Utility

Erica Guttman, Environmental Program Planner, Rhode Island Solid Waste Management Corporation

Howard Levenson (ex-officio), Senior Analyst, Office of Technology Assessment, U.S. Congress

Odonna Mathews, Vice President, Consumer Affairs, Giant Food, Inc.

Janet Matthews, Legislative Director, New York Legislative Commission on Solid Waste Management

Beth Quay, Manager, Recycling Planning and Programs, Coca-Cola U.S.A.

Karen Rasmussen, Senior Programs Manager, Corporate Environmental Programs, General Electric Company

Tom Rattray, Associate Director, Corporate Packaging Development, The Procter & Gamble Company

Paul Relis, Executive Director, Gildea Resource Center, Santa Barbara, California

Susan Selke, Associate Professor, School of Packaging, Michigan State University

Bruce Weddle (ex-officio), Director, Municipal and Industrial Solid Waste Division, U.S. Environmental Protection Agency

Wayne Wegner, Manager of Research and Development, Presto Products Company

Jeanne Wirka, Policy Analyst, Environmental Action Foundation

Alternates for Steering Committee*

Ed Fox, Associate Director, Corporate Packaging Development, The Procter & Gamble Company

Scott Fritschel, Development Programs Manager, Polymer Products Department, E. I. du Pont de Nemours & Company

Karen Hurst, Source Reduction Specialist, Gildea Resource Center, Santa Barbara, California

Peter Larkin, Director, State Government Relations and Environmental Affairs, Food Marketing Institute

Bob Weis, Manager, Plastic Waste Solutions, E. I. du Pont de Nemours & Company

Carl Woestendiek, Waste Reduction Planner, Seattle Solid Waste Utility.

World Wildlife Fund & The Conservation Foundation Staff

Gail Bingham, Vice President
Frances H. Irwin, Senior Associate
Christine Ervin, Associate
Sharon Green, Associate
Leah Haygood, Associate
Betsy Lyons, Research Fellow
William Silagi, Research Fellow
Eric Hostetler, Research Assistant
Blair Bower, Senior Fellow
Joy Patterson, Support Staff
Jenny Billet, Support Staff

Municipal Solid Waste Program, U.S. Environmental Protection Agency

Mike Flynn
Paul Kaldjian
Susan Mooney
Susan O'Keefe
Lynda Wynn

*In addition, our appreciation to David Buckner, formerly with the Illinois Department of Energy and Natural Resources, who participated as a Steering Committee member for the first half of the project.

Executive Summary

How can the design and use of products be altered to reduce the amount and toxicity of municipal solid waste? World Wildlife Fund & The Conservation Foundation, with a grant from the U.S. Environmental Protection Agency, launched a research project in 1989 to explore this fundamental question. It did so under the direction of a broad-based Steering Committee, whose efforts culminated in this final report, *Getting at the Source: Strategies for Reducing Municipal Solid Waste*. The report examines the evolving concept of source reduction, lays out an evaluation framework that decision makers in both the private and public sectors can use to help devise effective source reduction strategies, and takes a closer look at a few of those strategies, such as labeling programs for consumer products.

THE FOCUS IS ON PRODUCTS

Municipal solid waste is not simply amorphous quantities of garbage that must be recycled, incinerated, or buried. Garbage actually represents a stream of distinct products and materials we use every day as individual and institutional consumers. The various problems caused by products today—both in the ways we make them and use them—cannot be easily fixed by traditional approaches and add-on solutions. Creative, new approaches are needed to get at the source of our garbage problems by designing, making, and using products differently.

TOO MUCH STUFF

Americans are producing *too much stuff*. Each day, the average person generates about four pounds of municipal solid waste. That is nearly double the rate generated 30 years ago, and recent estimates for the U.S. Environmental Protection Agency (EPA) predict further increases in the decades ahead. There are concerns, too, that we frequently produce *the wrong stuff*. Some products and materials contain toxic substances that pose potential risks to human health and the environment; some generate potentially harmful by-products during manufacture; still others needlessly waste natural resources.

These problems cannot be eliminated by after-the-fact management practices alone—although recycling, in particular, must play an increasingly important role in managing wastes in the future.

- Managing wastes already produced, while vital to overall strategies for dealing with municipal solid waste, does not address underlying problems of waste generation.
- All waste management practices, including recycling, have associated economic and environmental costs, and can simply shift pollution problems from one environmental medium to another. For example, incinerating municipal waste produces air emissions and ash residues that require management and disposal on the land.
- Discarded products and materials in the municipal solid waste stream often represent resources that might have been used more efficiently elsewhere.

THE HIGHEST PRIORITY: SOURCE REDUCTION

In the face of such challenges and long-term risks, many have reached a simple but powerful conclusion: it is far better to reduce wastes in the first place than to cope inadequately with their aftermath.

> **By reducing the amounts and toxicity of waste generated, a wide range of benefits can be realized. Environmental impacts and potential risks can be diminished. Costs and handling problems associated with managing wastes also can be reduced. Finally, savings from using**

resources more efficiently can be realized by manufacturers and consumers alike.

AN OLD CONCEPT WITH NEW IMPERATIVE

Source reduction is not an entirely new concept. Some types of source reduction have been practiced over the years under different names—for example, "resource conservation" and "waste minimization." Manufacturers, for example, have long taken actions to reduce certain kinds of process wastes and to reduce the use of some materials. In response to prevailing economic incentives to reduce wastes, manufacturers have thus been able to make their operations more efficient and to help stabilize consumer prices.

Nevertheless, Americans continue to generate more and more waste—some of which is causing increasingly serious problems in the environment. Previous levels of effort are no longer sufficient to deal with those problems.

Why haven't we done more? The obstacles are numerous. For example, many people do not understand what source reduction really is, and often confuse it with recycling. In addition, there are few economic incentives in place that would encourage the level of source reduction needed. Individual consumers and manufacturers are rarely charged for the amount or type of garbage they produce; therefore, price incentives do not properly allocate resources in a way that would minimize the amounts and toxicity of wastes generated. Source reduction also involves fundamental changes in attitudes and behavior—changes that often run counter to prevailing lifestyle trends.

Source reduction is an old concept with new imperative. Today's problems are larger and more difficult. Existing economic incentives to address those problems are not strong enough, nor do they fully stimulate the degree of resource conservation needed today. Creative, new approaches are called for to identify and evaluate opportunities for source reduction.

GETTING FROM THEORY TO PRACTICE

The rationale for source reduction is compelling enough to cause many to cite it as the most important strategy for improving environmental quality in this country. Without concerted efforts to make the concept more tangible, however, the actual practice of source reduction will continue to lag behind its potential. *Getting at the Source* is intended to help fill in that gap by addressing various obstacles and offering recommendations to overcome those obstacles.

Steering Committee Recommendations

A NATIONAL DEFINITION FOR SOURCE REDUCTION

Even though source reduction is increasingly advocated as the highest priority approach to municipal waste problems, its meaning and application remain elusive. What exactly is a source? What types of wastes are to be reduced? When is source reduction achieved? No doubt the very richness in potential applications of source reduction has contributed to the profusion of alternative definitions currently in use. But it is important to resolve the current debate on definitions and move forward with analysis and actions. The Steering Committee recommends widespread adoption of the following definition:

Municipal solid waste source reduction is the design, manufacture, purchase, or use of materials or products (including packages) to reduce their amount or toxicity before they enter the municipal solid waste stream. Because it is intended to reduce pollution and conserve resources, source reduction should not increase the net amount or toxicity of wastes generated throughout the life of the product.

Several features of the definition are noteworthy. As stressed repeatedly throughout this report, source reduction involves multiple actors (not just manufacturers) and multiple activities (not just technical options). Second, it acts to conserve resources as well as to reduce pollution. Finally, source reduction extends beyond the municipal solid waste stream to the entire life cycle of products and materials.

Two broad approaches can be used to accomplish source reduction: through *design and manufacture* of a product (for example, using less material or eliminating toxic substances) and through *behavior* (for example, purchasing products that achieve source reduction or using products more efficiently). Numerous options and examples of their application are detailed in chapter 2.

CONSIDERING THE WHOLE LIFE CYCLE OF PRODUCTS

By definition, source reduction should not increase the net amount or toxicity of wastes generated throughout the life of a product. This provision is intended to ensure that efforts to reduce wastes in one part of the waste stream, such as municipal solid waste, do not produce an increase in the amount or toxicity of wastes elsewhere. A central question posed for decision makers, therefore, is whether or not tools exist to identify and evaluate net changes in wastes over a product's life cycle.

Studies that identify life-cycle characteristics of a product (known under a variety of names including "cradle-to-grave" studies) have been conducted over the past 25 years or so. A typical study identifies energy use, material inputs, and wastes generated during a product's life: from extraction and processing of raw materials, to manufacture and transport of a product to the marketplace, and, finally, to use and disposal of the product.

Information typically has been used for internal decision making by manufacturers. Recently, however, applications have extended into the public arena for such uses as consumer marketing and product labeling. That trend has prompted considerable debate about life-cycle studies, their policy implications, and underlying methodological issues. To explore such issues, various Steering Committee members participated in a forum convened by World Wildlife Fund & The Conservation Foundation in May 1990, and a workshop organized by the Society of Environmental Toxicology and Chemistry in August 1990, dealing with life-cycle studies.

Among other recommendations, the Steering Committee suggests that:

- **A three-part Life-Cycle Assessment model should be widely adopted. That model consists of: (1) an *inventory* of materials and energy used, and environmental releases (air, water, and land) from all stages in the life of a product or process; (2) an *analysis of potential environmental effects* related to the use of energy and material resources and environmental releases; and (3) an *analysis of the changes needed* to bring about environmental improvements for the product or process under study.**
- **Most, if not all, studies prepared to date have focused on the inventory component of the model. But inventories should not be used as the sole basis for drawing conclusions concerning the relative environmental effects associated with alternative products or materials.**
- **Unlike procedures used for most existing studies, inventory data on releases to the environment should be presented as an itemized listing of their types and amounts. This approach minimizes problems of implicit weighting of environmental effects and preserves the detail needed for successive analysis of effects.**
- **Peer review mechanisms and use of focus groups should be used in life-cycle assessments—particularly those to be used in the public arena—to help ensure the accuracy of information produced as well as a check on the quality of its presentation.**
- **While considerable progress is being made to define acceptable procedures for the inventory component of the Life-Cycle Assessment model, follow-up efforts are needed to explore and implement the remaining two components of the model.**

SETTING GOALS AND MEASURING PERFORMANCE

Municipal-solid-waste source reduction needs to be recognized—and implemented—as a national priority. Quantifiable goals would help stimulate and shape its development; measurement techniques would allow the tracking of its progress. Without this dual process, source reduction stands little chance of reaching its full potential.

Setting national goals is not, however, without significant challenges. Decision makers face an apparent dilemma between the need for accurate data and the need for action. Insufficient data and measurement tools exist upon which to base and track quantifiable goals; however, the very act of setting such goals can stimulate the types of activities needed to generate sufficient data and measurement tools. This dilemma is compounded by the fact that it can be inherently difficult to measure waste that has been prevented. In other words, how does one measure the amount of waste that has *not* been produced?

To address these types of challenges, the Steering Committee recommends development of two different types of goals, aimed at different purposes and different audiences. The first is a *national policy goal* that serves to inspire—rather than dictate—implementation plans by others. It should serve as the impetus for developing the second type of goal—*quantifiable goals for specific sectors of the economy and for in-*

dividual organizations. Among other recommendations, the Committee suggests that:

- **EPA should set a national policy goal for reducing the amount and toxicity of municipal solid waste based on the best information currently available. Such a goal should set the overall policy direction for source reduction but should be seen as a target toward which to aspire, not as a uniform measure against which all sectors should be held accountable.**
- **EPA should also devise explicit mechanisms and measurement tools for generating the additional information needed to account for changes in the municipal waste stream attributable to source reduction, to report on progress toward source reduction goals, and to revise its national policy goal.**
- **Where information is available, quantifiable goals should be developed for the various socioeconomic sectors of society (for example, a state, region, or commercial sector) or a major component of the waste stream (such as nondurable goods or yard wastes).**
- **State governments should set quantifiable goals for the institutional waste streams they produce, as well as for the waste streams within their jurisdictions, and should measure performance over time. Likewise, individual businesses should set and monitor source reduction goals for their products and operations as part of their overall pollution prevention plans.**

A FRAMEWORK FOR EVALUATING SOURCE REDUCTION OPPORTUNITIES

Some of today's most hotly debated issues about products and their use concern their relative effects on the environment. Which causes less of an impact, paper grocery bags or plastic? Cloth diapers or disposables? Paper plates or china? These are just a sample of the questions commonly faced by decision makers ranging from product designers and manufacturers to consumers and government officials. The underlying issue, however, is this: how can one tell if a change in the design and use of a product—or a change in a manufacturing process or activity—will actually result in less overall waste, or less toxic waste?

Answers to this question can be extremely complex and hindered by various obstacles. Often there is not enough information available to evaluate choices. But even where information is not a problem, there are numerous other factors that are difficult to evaluate, such as possible trade-offs in product performance or price.

A pervasive obstacle is the lack of guidance on *how* one should go about identifying and evaluating such factors in the first place. Where does one start? How can analysts systematically progress through the maze of possible choices and know they are asking the right questions along the way? Chapter 2's Evaluation Framework addresses this obstacle.

The Steering Committee recognizes that comprehensive analysis is needed to bring about meaningful reductions in the waste stream. To help decision makers in the public and private sectors evaluate source reduction opportunities, an Evaluation Framework is provided to guide and stimulate the creative approaches needed to get at the source of our waste generation problems.

The Evaluation Framework consists of a five-step decision-making process:

- *Step 1 — What is the problem?*
- *Step 2 — Which source reduction activities (or "options") should be considered further?*
- *Step 3 — Based on a wide range of considerations, which options promise to be the most effective?*
- *Step 4 — Which implementation strategies might be considered to put the source reduction option into action?*
- *Step 5 — Which of those implementation strategies promise to be most effective?*

To work through those steps, the Evaluation Framework presents a series of checklists and sets of questions to be used by analysts in ways that are most appropriate for their individual interests and needs. (For example, the checklist used in Step 2 presents a list of 30 different types of options that manufacturers and consumers might apply to a particular product, and gives examples of their applications.)

The Evaluation Framework offered by the Steering Committee is intended to be a source of ideas and to suggest factors that one may have overlooked. It is meant to stimulate creativity in identifying and evaluating source reduction opportunities, not to prescribe a rigid formula.

STRATEGIES WE CAN IMPLEMENT TODAY

Numerous types of source reduction strategies can be pursued without the need for extensive research. The

Steering Committee points to examples of strategies that are working already for municipalities, businesses, schools, and individuals. These include:

- **eliminating at least 10 percent of a city's residential yard waste by encouraging individuals to compost grass clippings and leaves in their own back yards;**
- **creating incentives for reducing community waste by charging households according to how much garbage they set out for collection;**
- **practicing source reduction at the office by copying on both sides of paper and using durable dishware;**
- **changing procurement practices by purchasing paper and inks that are nontoxic and buying equipment that will last longer;**
- **conducting waste audits in businesses and industrial plants to identify ways to reduce wastes and inefficiency;**
- **educating design engineers, students, and consumers to foster tomorrow's opportunities for source reduction; and**
- **reducing waste from unsolicited mail by removing one's name from mailing lists.**

A NATIONAL AWARDS PROGRAM FOR SOURCE REDUCTION

Recognition of source reduction achievements for businesses and other organizations could raise public awareness of opportunities and provide additional incentives for innovation.

The Steering Committee recommends that an annual national awards program be established to recognize outstanding achievements in municipal solid waste source reduction. Recipients could include manufacturers, commercial enterprises, public interest groups, state and local government agencies, and educational institutions.

Among recommended characteristics of such a program, the committee suggests that selection criteria include: environmental merit, innovativeness, transferability, economic value, and commitment to environmental protection. Moreover, an independent panel of experts should select award winners, with technical assistance provided as needed by an independent testing organization.

PROGRAMS FOR LABELING CONSUMER PRODUCTS

Numerous polls and surveys suggest that consumers are interested in buying products that are relatively better for the environment than alternatives. Labeling consumer products for their environmental attributes is one method that may be used to inform consumers and stimulate municipal solid waste source reduction. Chapter 3 identifies advantages and constraints associated with two types of labeling programs: "environmentally preferred" and "standard setting."

Environmentally preferred programs, as practiced in Germany and Canada and now developing in the United States, are voluntary (manufacturers can choose whether to apply), positive (only relatively superior products are distinguished by logos or seals), and based on specific technical criteria. These criteria can go beyond municipal solid waste concerns and include the amount of energy consumed and pollutants generated. Such programs can harness market incentives to stimulate source reduction and can reduce consumer confusion surrounding competing advertising. On the other hand, it is challenging to design programs that adequately assess environmental attributes and convey information properly to consumers.

The Steering Committee believes that labeling programs in the United States could be beneficial under the following conditions:

- **The program is national in scope.**
- **Product categories are selected to ensure that labels accurately reflect superior environmental performance.**
- **Criteria for awarding the label are based on life-cycle analysis.**
- **Criteria are set so as to stimulate technological innovation.**
- **Criteria are made public.**
- **Criteria are periodically reevaluated and strengthened.**
- **Consumers are educated about the meaning and limitations of the label.**
- **Government and industry are consulted and included in the program.**

A *standards setting* type of labeling program controls the use of certain terms, such as "recyclable" or "source-reduced," by defining terms and establishing conditions for their use in advertising and product labeling. In addition to advantages offered by environmentally preferred programs, standards setting can protect consumers and manufacturers alike against the use of inconsistent or inappropriate use of terms in the market. It may be difficult to arrive at uniform, agreed-upon standards, although efforts are under way in the United States to do just that.

The Steering Committee believes that both types of labeling programs provide a means of harnessing consumers for "green" products and verifying claims of environmental superiority. Each approach addresses somewhat different needs and has different strengths and weaknesses. Because standard setting can be managed to control the use of symbols and seals, the Steering Committee believes that such programs, if undertaken, should be carefully integrated to avoid a confusing proliferation of messages.

WHAT ELSE NEEDS TO HAPPEN?

Various types of research should be undertaken if source reduction is to reach its potential. Among other recommendations, the Steering Committee points to the following research priorities:

- **developing methods to measure the amounts of waste generated and the extent of source reduction accomplished;**
- **determining how to implement the remaining two stages of the Life-Cycle Assessment model—the analysis of potential environmental effects and the changes needed to help bring about needed improvements in products and processes;**
- **improving knowledge about the sociological and behavioral interactions that underlie waste generation patterns in the United States and in other countries;**
- **finding out how consumer attitudes and perceptions affect source reduction, including the effect of product labeling programs on both consumers and manufacturers;**
- **evaluating how source reduction works in practice, not just theory, including the potential and limitations of economic incentives and disincentives;**
- **studying the economic impacts of using reusable products instead of disposable ones;**
- **developing realistic and safer alternatives to the use of toxic substances in products; and**
- **evaluating whether a national planning process—similar to the one used for the energy crisis in the 1970s—could be useful for source reduction.**

Introduction

Getting at the source of pollution problems is frequently cited as the highest priority among strategies for improving environmental quality in the United States. Often called "source reduction" or "pollution prevention," the strategy describes an array of activities that manufacturers and consumers can take to reduce the amount and toxicity of wastes *before* they require management in the environment. Its intuitive appeal is familiar in the adage "an ounce of prevention is worth a pound of cure."

Nevertheless, a clear understanding of source reduction and its diverse applications remains elusive to many. Without concerted efforts to make the concept more tangible, the actual practice of source reduction will continue to lag behind its potential. This report is intended to help fill in the gap between the promise and reality of source reduction as it applies to the municipal solid waste stream.

Otherwise known as "garbage," municipal solid waste is more usefully viewed as a stream of products and materials used by individual and institutional consumers. In that light, *Getting at the Source* addresses the following central theme: *How can the design and use of products be altered to reduce the amount and toxicity of wastes that communities must manage?*

The report first examines the underlying motivations for source reduction and its evolution over time. Chapter 1 shows that managing wastes already produced—through recycling, incineration, and landfilling—does not address the underlying fact that Americans generate too much garbage in the first place. Moreover, management practices impose economic and environmental costs that can be reduced by prevention. Finally, the emerging emphasis on source reduction also speaks to the heightened appreciation for using our resources more efficiently in this throw-away culture.

Source reduction is not a new concept. It incorporates practices used by manufacturers over the years to make their operations as efficient as possible and to help stabilize consumer prices. But source reduction is taking on a greater imperative today as environmental problems become more pronounced and the need for resource conservation becomes more important. Various obstacles must be overcome in that process—several of which are addressed in chapter 1.

First, is the need for a clear, concise, and consistently used *definition of source reduction*. The Steering Committee directing this project offers such a definition and illustrates the major approaches to source reduction. Decision makers also need to consider the resource and pollution characteristics of products throughout their life cycles of production, use, and disposal to avoid unintentionally increasing waste elsewhere. Accordingly, chapter 1 also offers recommendations concerning the challenges and opportunities for using *Life-Cycle Assessments* in the analysis of source reduction. A third obstacle is the lack of comprehensive goals for accomplishing source reduction in this country and difficulty in measuring accomplishments. The Steering Committee recommends *a goal-setting process* that addresses the needs and responsibilities for the various parties involved in source reduction.

Another major need addressed by the Steering Committee is the lack of pragmatic guidance available to decision makers to help identify and evaluate source reduction opportunities. How can one tell if a change in the design and use of a product—or a change in a manufacturing process or activity—will actually result in less waste? Where does a decision maker start? What are the right questions to ask, and how does an analyst sift through the maze of possibilities in a systematic and efficient way? Chapter 2's *Evaluation Framework* is designed to help provide that guidance. It consists of a five-step series of checklists and questions designed to stimulate thinking, highlight information needs, organize data, and help in making evaluative decisions. Examples from product categories, such as third-class mail and household batteries, are used to help illustrate how the Framework can be used.

Clearly, much research and evaluation needs to be undertaken by decision makers in both the public and private sectors to further the progress of source reduction. But many activities can be put into practice today. Chapter 3 takes a closer look at a few of those strategies already being used to great effect by individuals, manufacturers, and public institutions. The chapter also explores and offers recommendations for a national awards program and alternative labeling programs that aim to reward source reduction achievements. Finally, recommendations for needed research are identified to help ensure that source reduction becomes predominant in practice—as well as in theory.

Chapter 1
The Emergence of Source Reduction

A common theme of this report is that source reduction—reducing the amount and toxicity of wastes generated—will require fundamental changes in how we value and use our resources. That, in turn, will require changes in attitudes and behavior on the part of governmental and corporate entities to individual consumers. Such changes cannot occur without a better understanding of what source reduction is, why we need it, and how it can be applied in diverse settings by a wide range of actors.

This chapter explores various problems and trends associated with American garbage, challenges in the way we manage that garbage, and factors that might help explain patterns and trends. The chapter next analyzes the past achievements of source reduction and explains why it must assume a greater role today. To help shape that role, the Steering Committee proposes a national definition for source reduction, examines the potential use of life-cycle studies to make sure that wastes aren't simply transferred from one waste stream to another, and lays out recommendations for getting on with setting source reduction goals and measuring progress toward those goals.

DIMENSIONS OF THE GARBAGE PROBLEM

Americans are producing too much garbage.* Costs of managing that garbage are escalating as landfill capacity declines and new, improved facilities become more expensive. Furthermore, there are increasing concerns about the kinds of garbage produced. While many view municipal solid waste as amorphous quantities of garbage, it actually represents a stream of distinct products and materials we use every day as individual and institutional consumers. Some of those items contain toxic substances that pose potential risks to human health and the environment; some generate potentially harmful by-products during the course of manufacture; still others appear to be needlessly wasteful of the resources required to manufacture and use those goods.

Recycling, incineration, and landfilling will continue to be critical elements of the overall strategy needed to handle municipal solid wastes. But those management practices can impose significant financial and environmental costs to society and do not address the underlying problems of waste generation. Such problems, as discussed below, point to the need for source reduction to reduce the amount and toxicity of wastes.

Too Much Stuff

Americans are generating more garbage than ever before. The overall amount of municipal solid waste doubled in less than three decades—from 88 to 180 million tons between 1960 and 1988.[1] Between 1985 and 1988 alone, the amount of waste generated in this country increased by nearly 20 million tons. That represented an 11 percent increase in just three years! And projections show an additional 39 percent will be generated by the year 2010.

Only about half of that historical increase can be attributed to growth in the population base alone. The U.S. Environmental Protection Agency (EPA) estimates that per capita generation rates have increased from about 2.7 pounds per day in 1960 to 4.0 pounds in 1988—nearly a 50 percent increase per person.

*Municipal solid waste or "garbage" generally refers to solid wastes from residential, commercial, institutional, and certain industrial sources. It includes such items as durable goods (for example, major appliances and furniture), nondurable goods (for example, newspapers, office papers, clothing), packaging, food, and yard wastes. The term typically excludes such wastes as municipal sludges, combustion ash, and industrial process wastes that might also be disposed of in municipal waste landfills.

Figure 1.1
Total and Per Capita Generation of Municipal Solid Waste (by weight)

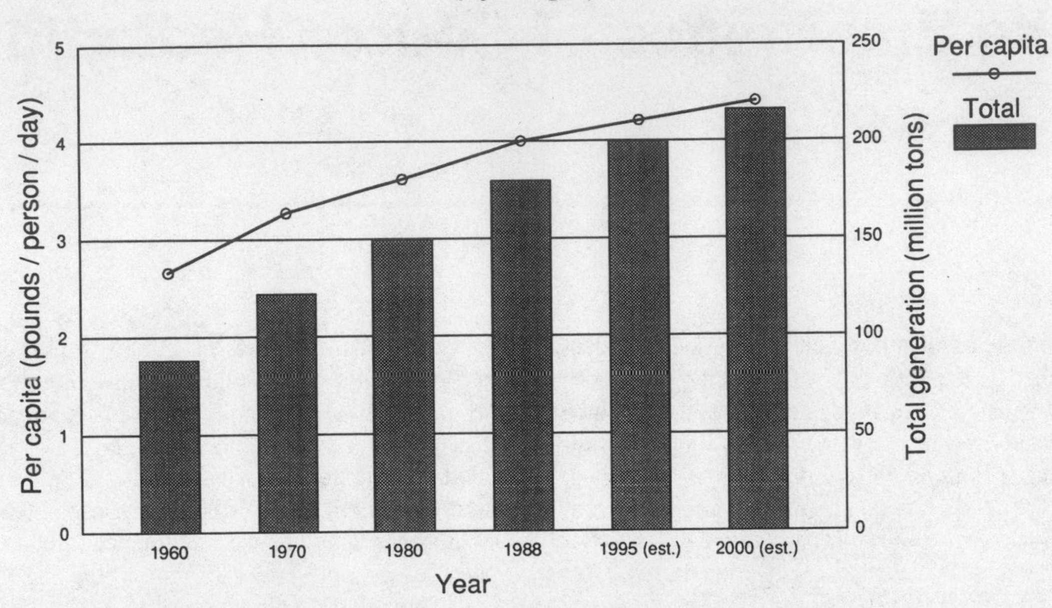

Source: U.S. Environmental Protection Agency.

Projections show that rate approaching 4.9 pounds per day in the year 2010.

These estimates may be low. A survey of about 40 cities and counties showed that per capita generation rates ranged from about 2 to 9 pounds per day, although these figures also reflect differences in the way localities define and measure municipal solid waste.[2]

The relationships among per capita waste generation, economic activity and cultural differences are not straightforward. There are indications that Americans generate more waste per capita than residents in certain industrialized countries, such as Sweden and West Germany. One recent analysis of international data, for example, found no significant correlation between the per capita amount of municipal solid waste generated and per capita income across 24 countries.[3] Most researchers agree that considerably more research is needed to be able to compare and interpret international data due to differences in the way municipal solid waste is defined and measured.[4]

The Wrong Stuff

Concerns over how *much* waste is produced are accompanied by concerns over what kind of waste is produced. Toxic materials in the waste stream increase the potential risks to human health and the environment associated with incineration, recycling residues, and groundwater contamination from landfill leachate.[5] Some of those materials enter the waste stream through products recognized as household hazardous waste (for example, batteries, paints, pesticides, and herbicides). Other common household products may contain potentially toxic materials such as residues from certain types of printed materials and pigmented plastics. Still other sources include certain industrial and commercial firms that are allowed to deposit small amounts of hazardous wastes in municipal landfills or have them burned in municipal incinerators.[6] Nevertheless, very little information exists on the distribution of toxic substances in the waste stream and what risks are associated with their presence.

Other components of the waste stream prompt questions about our throw-away culture that transcend concerns about the quantity of waste generated alone. Certainly the packaging component of the waste stream has attracted considerable attention across the country. As discussed in a later section, the total amount of packaging in the waste stream has more

Figure 1.2
Changing Strategies for Dealing with Municipal Solid Waste

	Actual 1960	Actual 1988	Projected 1995*
Landfilling	62%	73%	53%
Material Recovery	7%	13%	24%
Incineration	30%	14%	23%
Reduce generation (Source reduction)	—	—	—

*1995 projections reflect midpoints—for example, recycling and composting ranges from 20 to 28 percent.

Source: U.S. Environmental Protection Agency.

than doubled in the past 30 years. Products introduced in "convenience" packages were projected to increase 10 percent in 1990 alone according to one estimate.[7] When nondurable products (for example, newspapers, magazines, telephone books, third-class mail) are added to packaging items, more than one-half of the total municipal solid waste generated in the United States consists of products and packages with life spans that are often quite short. Third-class mail, much of which is thrown away without being opened, comprises about 1.5 percent of the entire waste stream. A recent survey in Seattle found that 62 percent of the respondents said they would like to have their name removed from mailing lists but that only 17 percent had taken the actions necessary to do so.[8]*

Factors underlying the trend toward more disposable products are complex and sometimes involve trade-offs among different objectives. Some types of packaging, for example, have been added to protect consumer safety as well as to provide convenience. Joint private-public efforts are under way in the Northeast region, and in states such as Minnesota and Washington, to find ways to identify and eliminate packaging that is considered to be unnecessary.

After-the-Fact Management Is Not Enough

Dealing with municipal solid waste problems requires action on a variety of fronts including source reduction and improved methods of waste management.† Regardless of the encouraging developments in management practices discussed below, waste management can be costly and poses various environmental impacts. Treating wastes at the "end-of-the-pipe" can also simply transfer pollutants from one medium to another. For example, ash residues from the incineration of municipal waste pose treatment and management challenges in their own right. Figure 1.2 shows how waste management practices are changing in response to these types of problems.

Landfilling

Much of the recent publicity about garbage problems has focused on landfill closures and the controversial siting of new facilities. About 14,000 landfills were operating in the United States in 1978. Based on the latest survey, 45 percent of the nation's 6,000 existing landfills were expected to reach their currently authorized capacity by 1991.[9] Others are expected to close rather than attempt to meet stricter federal standards under consideration.

Information on the number of closures, however, does not easily translate into estimates of capacity. Some landfills are being expanded while others are being replaced with larger facilities.‡ Where storage capacity is at a premium—in the Northeast and portions of the Midwest, in particular—disposal costs have escalated sharply. The average fee charged to waste haulers to unload at landfills (called "tipping" fees) in the Northeast region more than doubled between 1986 and 1988—from $21 to $45 per ton. These fees were at least two-and-one-half times higher

*See chapter 3 for an explanation of what steps are needed to remove one's name from unwanted mailing lists.

†This report does not address the various methods and issues associated with composting municipal solid waste other than backyard composting, which is a form of source reduction (see chapter 3).

‡Nationwide estimates of overall capacity are hindered by incomplete and/or inconsistent[10] methods of reporting landfill closures and expansions.

than any other region.[11] Some waste haulers have responded by "exporting" their wastes to lower-cost regions and states for disposal. Beyond these critical areas, however, landfills are becoming an increasingly expensive option for communities due to the rising costs of construction and operation for landfills that must meet improved environmental standards.

Many landfill closures reflect the progress made in eliminating facilities causing environmental impacts. Thousands of open dumps—often noted for their unsanitary conditions, methane explosions, and releases of hazardous substances to the environment—have been eliminated. Indeed, 20 percent (about 250) of the sites on the Superfund National Priorities List as of 1988 were municipal landfills.[12] While the current generation of landfills and attendant regulatory controls are major improvements, debate continues over the relative merits and potential risks associated with alternative landfill designs.

Incineration

The role of incineration has changed significantly over the past three decades. In 1960, about 30 percent of municipal solid waste was burned in facilities that lacked pollution control devices; many also lacked capacity for energy recovery. As these facilities were shut down, the share of municipal waste handled at incinerators reached a low of 10 percent in 1980. That share is gradually increasing once again.

Today's generation of incinerators is greatly improved. Many facilities are now equipped with air pollution control and energy recovery technology.[13] Nevertheless, other facilities remain out of compliance with regulatory standards, and concerns persist over the toxicity of emissions and ash residues from incineration in general. There also is controversy about whether or not increased reliance on incineration will discourage source reduction and recycling efforts. Incineration greatly reduces—but does not eliminate—the amount of municipal solid waste needing to be landfilled. For example, efficient mass-burn facilities tend to reduce the amount of incoming municipal solid waste by about 80 to 90 percent by volume and about 65 to 75 percent by weight.[14] Resulting ash residues, when added to materials that cannot be burned (for example, major appliances, construction debris), and wastes that are bypassed to the landfill when an incinerator is undergoing repair and/or maintenance can amount to perhaps 40 percent of the landfill volume a community would need without an incinerator.[15]

Recycling

Recycling is the option preferred by EPA and many other groups for managing wastes already produced. Overall recycling rates have increased from about 7 percent of the waste stream in 1960 to about 13 percent in 1988. Measured against current EPA and state recycling goals, which range from 25 to 50 percent, the 1988 level is well below policy objectives. Nonetheless, improvements in the past couple years have been substantial. There were a total of 2,700 curbside recycling programs in 1990, up from about 1,000 just two years before.[16] Some cities, such as Seattle and Minneapolis, are reporting recycling rates exceeding 25 percent.[17] In addition, many state legislators are taking action to buttress recycling efforts by enacting "content" legislation that requires the use of designated amounts of secondary material in the manufacture of paper and other finished goods. By the end of 1989, a total of 125 recycling laws had been passed in 39 states and the District of Columbia addressing such issues as educational and assistance programs, tax credits, and certain mandatory provisions such as content requirements.[18]

While the progress of recycling is rapid, ability to meet EPA, state, and local recycling goals remains constrained. Key obstacles include the lack of reliable markets for secondary materials, costs in transporting materials to those markets, inadequate collection and separation capabilities, and the lag between setting goals and necessary investments in new facilities and retooling of old facilities.

Although the environmental impacts associated with recycling are believed to be less than with landfilling and incineration, recycling does produce pollution. Collection and processing recyclable materials produces air emissions, and the manufacturing of finished goods results in residues of a nonhazardous and hazardous nature. A systematic comparison of these impacts and those of landfilling and incineration has not been conducted. However, several studies suggest that recycling is by far environmentally preferable to use of virgin resources when the full cycle of processing, use, and disposal is taken into account.[19]

Trends in the Composition of Municipal Solid Waste

Getting at the source of our municipal solid waste problems requires accurate understanding of what materials and products are generated and discarded, by whom, and factors explaining those patterns. Such information is necessary not only for targeting areas of the waste stream for analysis of source reduction opportunities but also for devising effective policy strategies. A few of the most dominant trends are highlighted below and in figure 1.3.[20]

Changes in Product Composition

Trends in the mix of product categories making up the municipal waste stream generally depict a redistribution from food and yard wastes to nondurable and durable goods (see figure 1.3).

Containers and packaging comprised the single largest portion of wastes generated in 1988 (about 32 percent of the total in terms of weight). Packaging wastes more than doubled in the past three decades (from 27 to 57 million tons)—a slightly higher rate of increase than for the overall waste stream. However, packaging's rate of increase has dropped steadily since an immense surge between 1960 and 1970 (a 59 percent increase in 10 years), and its overall share of the waste stream is now declining very slowly. That downward trend does not suggest that the amount of packaging is declining; it is projected to increase by nearly 1 million tons each year over the 10 years or so. Rather, the trend reflects relative changes among product categories over time and the substitution of lighter-weight materials such as plastics and aluminum for the use of glass and steel.

Nondurable goods are projected to take over the lead as the single largest category of municipal solid waste by the turn of the century. Defined as goods with a lifespan of less than three years, nondurable products include such items as newspapers, books and magazines, office papers, advertising materials, disposable diapers, and clothing. From a percentage share of just 20 percent in 1960, it is expected to reach nearly 32 percent in the year 2000—an increase of nearly 1.5 million tons each year (see figure 1.3). Items showing the highest amount of increase are paper products—especially books and magazines, office papers, and commercial printing (advertising).

Wastes associated with *durable goods* (for example, appliances, tires, furniture and other longer-lasting products) will increase by about 500,000 tons per year over the next decade. Most of this increase, as well as the increase in percentage share of the waste stream, is associated with furniture and furnishings and "miscellaneous" durables such as small appliances and consumer electronics.

Yard wastes comprise about 18 percent of the total waste stream today (by weight). That relative share is expected to decline based on the EPA/Franklin assumption that per capita generation rates for yard wastes will stabilize after 1988. The per capita generation rate for food wastes is expected to decline slightly due to greater use of preprocessed foods, but also because greater use of garbage disposal systems simply transfer food wastes down the drain for treatment in the water system.

Changes in Material Composition

Paper and paperboard has comprised the largest share of the waste stream since about 1965 (see figure 1.3). From 72 million tons each year in 1988 (40 percent of the total), paper wastes are expected to increase to 96 million tons by the year 2000 (44.5 percent).

Changes among materials in the waste stream are more extreme than they are among major product categories. As a percentage share of the waste stream, *yard and food wastes* have declined from 37 percent in 1960 to 25 percent in 1988—a 40 percent decrease even as the per capita amount generated continues to increase slightly as discussed above. Perhaps the most conspicuous trend has been the relative shift from *glass and ferrous metals* to *plastics*, which were less than 1 percent of the waste stream in 1960. Of the 14 million tons of plastic products generated in 1988, about 40 percent is used for containers and packaging, 27 percent is used in miscellaneous nondurable goods such as trash bags and eating utensils, and much of the balance is used in durable goods.

Factors Influencing Generation Rates and Trends

Socioeconomic research generally is not available to help explain behavioral aspects underlying trends in our consumption and discard of goods. A brief discussion of implications from available research demonstrates how important this type of information can be in understanding the opportunities and challenges for reducing the generation of municipal solid waste.

Population

Overall population levels are the dominant factor underlying many of the quantity-based trends discussed above. Population growth alone, however, accounts for only about half of the twofold increase in the municipal solid waste generated in the past several decades. Projections indicate that average annual rates of increases for waste generated will be nearly double the rate of population growth over the years 2000 to 2010 as well.[21] It is important, therefore, to better understand some of the factors underlying trends in the increase of per capita generation rates.

Figure 1.3
Changes in Products and Materials in the Waste Stream
(percent by weight)

Major Product Categories

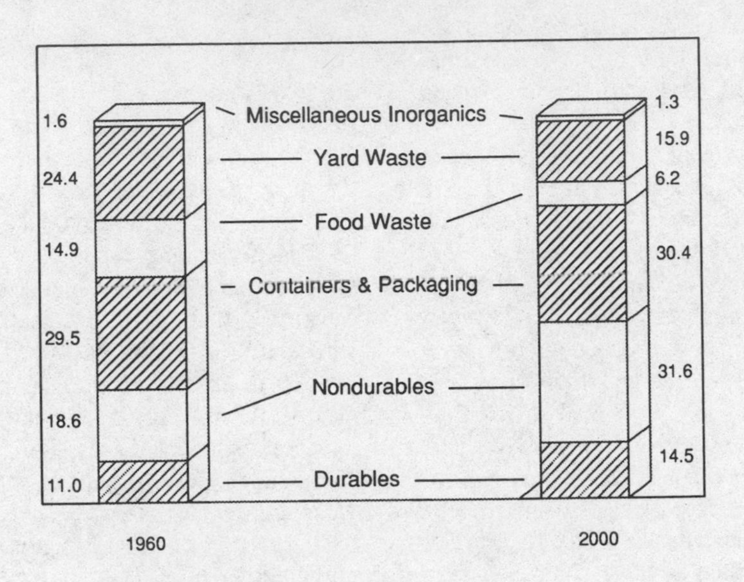

Materials

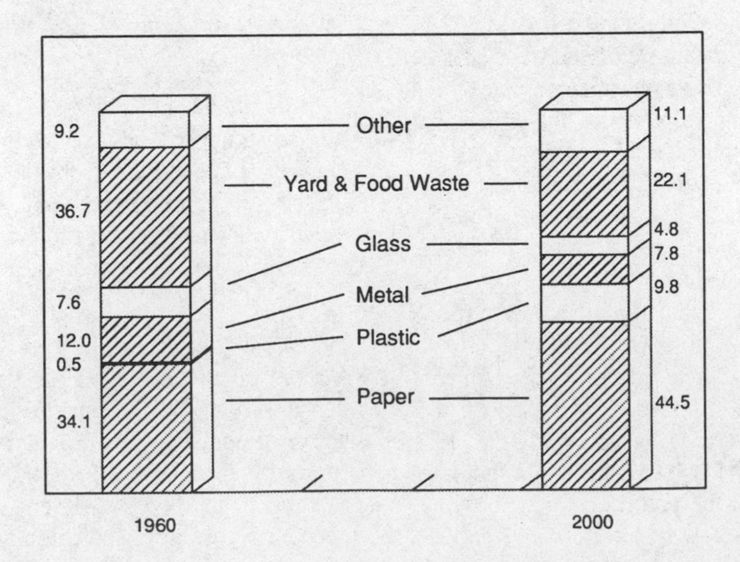

Source: U.S. Environmental Protection Agency.

Disposable Income

A recent study in Seattle bolsters the common assumption that growing affluence in the United States is associated with high rates of waste generation. In that study, residential household garbage (apart from items set out for recycling) increased about 0.6 percent for every 1 percent increase in household income.[22] On the national level, municipal solid waste generation increased at about two-thirds the rate of real disposable personal income over the period between 1970 and 1988.[23] Research on how income affects individual components of the waste stream, however, is scarce and inconclusive. Thus far, income is shown to be correlated with only a few components of the waste stream. For example, one study found that higher-income households produce more yard debris and newspapers while lower-income households tend to produce more packaging of all types.[24] Reasons for the latter may include a higher proportion of pre-prepared foods and beverages, and less cash flow and storage space to purchase products in bulk. Such socioeconomic differences, if found to be common, could be important in tailoring effective source reduction strategies to various groups.

Size of Household/Changes in Labor Force

Limited research suggests that per capita rates of waste generation increase as the size of households decreases.[25] This inverse relationship has some significance, of course, given the trend toward smaller households in this country. For example, the average household in the Seattle metropolitan area decreased from 2.52 persons in 1968 to 2.02 in 1987. Research in that city confirms that the average amount of residential waste generated in smaller households is proportionately higher than that produced by larger households.[26] Underlying factors for this type of relationship might include increased reliance on time-saving products and fewer opportunities to share goods, such as newspapers, among household members. It might also be assumed that the increasing number of women in the labor force also results in similar preferences for time-saving products, but research does not appear to be available to explore such relationships.

Population Density/Degree of Urbanization

One recent study found that per capita generation rates are lower in rural areas—at least for certain products such as newspapers.[27] A recent study in Minnesota also noted that overall per capita rates were significantly higher in the Twin Cities metropolitan region compared with the rest of the state, but findings here and elsewhere can sometimes reflect different assumptions for measurement.[28] Additional research on this type of relationship could help target certain source reduction measures.

Geographic and Seasonal Variations

Local studies demonstrate that the overall amounts and composition of municipal wastes vary significantly depending on location and season.[29] For example, yard wastes make up a much greater share of the municipal waste stream in the summer and fall seasons, while various paper products associated with packaging can peak during the winter holiday season. This type of variation underscores the need for local waste composition studies.

Disposal Costs

One of the greatest challenges for source reduction stems from the fact that generators of municipal solid waste—whether they be individuals, businesses, or manufacturers—rarely are charged for the full costs associated with disposing or otherwise managing that waste. Individual consumers, in particular, rarely pay for garbage disposal directly but through general taxes. Unlike many other transactions in the marketplace, therefore, product demand (in this case, the amount of waste generated through purchasing and discarding products and materials) is not moderated by price incentives. Very little research has been done to explore that relationship with respect to municipal solid waste, but general experience with economic incentives suggests that this may be one of the most promising areas for testing and analysis.*

Convenience and Other Lifestyle Factors

Perhaps the most difficult factor to assess concerns general lifestyle choices. To what extent do apparent trends toward convenience products, disposability, and the expanding variety in models, sizes, and colors of products affect the overall amounts and toxicity of waste generated? How are these trends shaped by factors such as consumer attitudes, prices that do not reflect environmental costs, and advertising? How should source reduction strategies address such trends?

Single-use or disposable products, for example, often cost less than reusable alternatives in the short

*See chapter 3 for a discussion of how economic incentives are being employed in residential collection programs through variable can rates.

run, and are more convenient and/or less time-consuming. A recent survey found that 93 percent of Seattle's residents said they preferred long-lasting to disposable products; 42 percent, however, said they liked to use disposable products.[30] Very little research has been reported that would lend insight to how information on short- versus long-run costs would affect such consumer choices. Nor is it understood how attitudes and marketing practices in the United States differ from those in other developed countries that may not exhibit as high a proportion of "convenience" products.

Many people intuitively regard the trend in product fads to be wasteful of resources, but research does not appear to exist that would document the extent of this trend or to explore the underlying cultural and economic incentives. Sometimes apparently minor changes in products, such as adding a color or fragrance, can have other implications for waste generation. Studies of manufacturing processes, for example, demonstrate that each step added to a production sequence tends to increase the amount of waste generated in the manufacturing process.[31] While such life-cycle wastes may not necessarily wind up in the municipal solid waste stream, they are important considerations in evaluating source reduction opportunities.

THE HIGHEST PRIORITY: SOURCE REDUCTION

As the preceding section indicates, there are various motivations for reducing the amount and toxicity of municipal solid waste:

- Managing wastes already produced, while vital to overall strategies for dealing with municipal solid waste, does not address underlying problems of waste generation.
- All waste management practices, including recycling, have associated economic and environmental costs and can simply shift pollution problems from one environmental medium to another (for example, air, water, land).
- Discarded products and materials in the municipal solid waste stream often represent resources that might have been used more efficiently elsewhere.

In the face of such challenges and long-term risks, many have reached a simple but powerful conclusion: **Far better to reduce wastes in the first place than to cope inadequately with their aftermath.[32] By reducing the amounts and toxicity of waste generated, a wide range of benefits can be realized. Environmental impacts and potential risks can be diminished. Costs and handling problems associated with managing wastes also can be reduced. Finally, savings from using resources more efficiently can be realized by manufacturers and consumers alike.**

Why the Delay in Emphasizing Source Reduction?

Given these compelling motivations and advantages, it is reasonable to question why source reduction has not emerged previously as the first choice in dealing with municipal solid waste problems. There are numerous reasons. First, many people do not yet understand the concept of source reduction and often confuse it with recycling. Second, source reduction involves fundamental changes in attitudes and behavior. Those changes will have to be pervasive throughout society, affecting people ranging from individual consumers and policy makers to product designers and manufacturers. Shifts in attitudes to embody the principles of source reduction often run counter to our throwaway culture and trends that place a premium on convenience and style.

Third, there are currently few economic incentives for reducing the amount and toxicity of municipal solid waste. Unlike the charges assessed on manufacturers for the disposal of hazardous wastes, which provide direct economic incentives to reduce those wastes, manufacturers are *not* charged for the costs of disposing products in the municipal waste stream. Likewise, the prices paid for most consumer goods do not reflect the full range of environmental costs associated with their manufacture, use, and disposal. Resulting prices are lower than they would otherwise be and demand is correspondingly higher. In addition to proper price "signals," adequate information on market choices is important for the efficient operation of free markets. Many people, however, do not have sufficient information for selecting products based on their relative environmental impacts.

Finally, there are inevitable lags between changes in attitudes and corresponding changes in the underlying institutional and economic system that responds to consumer demands. Whether or not those attitudes are really changing, and to what extent, is still subject to debate.

Understanding the obstacles to source reduction is not always straightforward. For example, some trends that appear to work against source reduction—such as increased packaging for certain products—have been motivated by the need to protect consumer and worker safety and to reduce product tampering. Sometimes there is a gap between what consumers express as a preference (for example, through public opinion polls) and their actual behavior in the marketplace. Moreover, it is difficult to determine how consumer demand is affected by various factors such as product prices and marketing techniques. These types of considerations and trade-offs must be evaluated when exploring the potential for source reduction.

An Old Concept with New Imperative

Like many "new" concepts, source reduction draws from the past but is evolving in response to new needs and circumstances. Certain types of source reduction have been practiced over the years under such names as "resource conservation" or "waste reduction." Manufacturers have long taken measures to reduce their use of materials in order to operate more efficiently and to stabilize consumer prices in the face of inflation.* Economic incentives also have stimulated the reduction of process wastes.

Source reduction—as a means to reduce costs in managing pollution and in using resources more efficiently—is seen by some corporate leaders as a tool to bolster competitiveness in global markets. With respect to wastes released to the environment (air, water, and solid wastes), incentives for reduction were relatively small until enactment of the major environmental legislation in the 1960s and 1970s. Those incentives have steadily increased in terms of stricter regulatory controls on wastes, increased liabilities, and public concern over environmental quality in general.

Despite the achievements made to date, factors discussed earlier have led to increases in per capita generation rates as well as increases in the overall amount of waste generated. Even though the level of waste has decreased per unit of output for some products, the increase in overall market production is producing significantly more waste with which to deal. Associated environmental problems are increasing and are sometimes becoming more complex as pollutants interact in ways that are not fully understood. Moreover, resource conservation is once more taking on greater importance.

Source reduction can be seen, therefore, to be taking on increased significance in response to today's environmental problems. Existing approaches for dealing with waste generation are no longer sufficient. Creative, new approaches are needed and stronger incentives provided to bridge the gap between the promise—and reality—of source reduction.

*In the packaging industry, for example, each metal can uses just 25 percent of the amount of metal used 30 years ago; plastic bottles use about 50 percent of the plastic; and glass bottles 75 percent of the glass formerly used—all in response to economic forces.

UNDERSTANDING SOURCE REDUCTION

Even though source reduction occupies the pinnacle of the municipal waste management hierarchy, its meaning and application remain elusive to many. What is the "source" to which the phrase refers? What is to be reduced? Defining source reduction is further complicated by the fact that it is unclear what the concept is. Is it a goal? A policy? A technique? Clearly, "source reduction" is best understood when placed in a specific context.

> *"Source reduction involves changes in the way we design and use products that will affect kitchens, boardrooms, drafting rooms, factory floors and legislative halls."*
>
> — Gildea Resource Center
> Santa Barbara, 1988[33]

What is the "source"? In the context of hazardous waste, source reduction has been defined by EPA as "any activity that reduces or eliminates the generation of hazardous waste at the source, usually within a process." This definition presumes that the source of waste is an industrial plant. In the context of municipal solid waste, the sources are diverse. Wholesalers, retailers, hospitals, public and private offices, municipal operations, residences, and some industries, among other generators, can be considered sources of municipal solid waste. In addition, there are multiple sources associated with many kinds of products and packaging (for example, solid and other wastes may be produced at various stages of the manufacture, distribution, and use of a product.)

What types of "wastes" are to be reduced? It is important to consider which wastes are to be the target of reduction efforts because policies to reduce one category of waste are likely to affect the generation of others. Wastes can be characterized as follows:

- *Post-consumer wastes:* solid wastes generated by all end-users (including individuals, institutions, governmental agencies, and manufacturing plants) from materials that have served or are unsuitable for their original intended use.
- *Life-cycle wastes:* materials and energy wastes

Figure 1.4
Approaches to Source Reduction

DESIGN AND MANUFACTURE: Manufacturers can modify the design and manufacture of products and materials to:

Reduce the amount of waste by

- Eliminating or using less material to accomplish a given function (as long as any material substitution does not increase the net amount or toxicity of wastes throughout the life of the product);
- increasing the lifespan of the product; and/or
- Enhancing the reuse, repair, and remanufacturing of the product or its components

Reduce the toxicity of waste by

- Eliminating or minimizing the use of toxic substances in a product or process and/or
- Substituting nontoxic or less toxic substances

BEHAVIOR: Users (individual and institutional) of products or materials can modify their behavior to reduce waste by:

- Purchasing products that have been subject to source reduction by one of the means outlined above;
- Purchasing different products (for example, durable coffee mugs rather than disposable);
- Using products differently (for example, copying two-sided, reusing products); and
- Managing potential wastes before they enter the municipal solid waste stream (for example, composing yard wastes).

generated during the entire life of the product/package, including both preconsumer and postconsumer wastes. Sometimes referred to as "cradle-to-grave" wastes in the literature, these wastes stem from extraction of raw materials, manufacture, transportation, distribution, use/consumption, and disposal—and can be released to all media (land, air, water).

The relationship of these categories of wastes to the definition of municipal solid waste is not direct. While municipal solid waste clearly includes most postconsumer wastes, it may also include other life-cycle wastes, depending on how municipal solid waste is defined and managed in a particular community. It may include varying amounts of wastes generated by manufacturers and commercial enterprises in their roles as consumers but usually does not include wastes associated with the extraction and processing of raw materials.

Source Reduction: A Definition

Clearly, the "straightforward" concept of source reduction carries broad implications for the use and consumption of resources in our society. It touches the way in which Americans live, the products they consume, and how they use those products. This very richness in potential applications no doubt has contributed to the profusion of alternative definitions currently in use. It is important, however, to resolve the current debate over definitions to move ahead with needed analysis and implementation. The Strategies for Source Reduction Steering Committee recommends the following definition be adopted:

> **Municipal solid waste source reduction is the design, manufacture, purchase, or use of materials or products (including packages) to reduce their amount or toxicity before they enter the municipal solid waste stream. Because it is intended to reduce pollution and conserve resources, source reduction should not increase the net amount or toxicity of wastes generated throughout the life of the product.**

Examples of Source Reduction

Some examples of source reduction of municipal solid waste are relatively easy to identify. Reducing the amount of material used in a product or package that will enter the waste stream (without otherwise increasing life-cycle wastes) accomplishes source reduction. Making two-sided photocopies reduces waste. And, to the extent that yard wastes are composted or otherwise dealt with on site, they can be said to be reduced at the source. These and other examples are discussed further in chapters 2 and 3.

Other examples of source reduction are more difficult to evaluate. Suppose an office switches to the use of recycled-content paper that is slightly heavier than virgin paper used previously. The life-cycle wastes associated with the change may be reduced, but the amount of waste paper entering the waste stream is increased. Has source reduction been achieved? If a manufacturer switches from a recyclable packaging material to a lighter, but less recyclable material, has source reduction been achieved? Such examples point to the need for a systematic method for measuring source reduction as is discussed in the next section.

EVALUATING PRODUCTS OVER THEIR ENTIRE LIFE CYCLES

By definition, source reduction should not increase the net amount or toxicity of wastes generated throughout the life of a product. This provision is intended to ensure that efforts to reduce wastes in one part of the waste stream, such as municipal solid waste, do not produce an increase in the amount or toxicity of wastes elsewhere. A central question posed for decision makers, therefore, is whether or not tools exist to identify and evaluate net changes in wastes over a product's life cycle.

This section introduces some of the issues surrounding life-cycle analysis, recommends adoption of a Life-Cycle Assessment model to help guide future analyses, and explains how the model relates to the evaluation framework presented in the next chapter. Finally, other recommendations are offered to help make life-cycle information useful to those interested in source reduction.

Background

Studies that identify life-cycle characteristics of a product have been conducted in the United States and Europe over the past 25 years or so. A typical study of this genre* identifies energy use, material inputs, and wastes generated during a product's life: from extraction and processing of raw materials, to manufacture and transport of a product to the marketplace, and finally to use and disposal of the product.

The actual scope and methodology of life-cycle studies have varied according to the needs of clients and approaches taken by various consultants. Most studies have compared alternative materials for a given product or package use and have been used primarily by manufacturers for internal decision making. Because results were not often used in the public sector, major questions concerning methodology and assumptions tended to be worked out between the consultant and client.

Over the past year or so, demand for life-cycle information has surged in tandem with renewed concern over environmental quality. One major consulting firm, for example, commonly performed about one study per year over the past two decades, whereas more than a dozen were in progress during 1990 alone. An increasing number of firms are moving into the field of life-cycle analysis in response to the increased demand. Perhaps the most significant policy development is the growing interest in using life-cycle information in the public domain, such as for consumer marketing, product labeling, and government policy formation as it might affect certain materials and products.

A Forum: Policy Issues and Implications of Life-Cycle Studies

The expanding array of applications brings life-cycle studies into the public arena as never before. As a result, there is a need to evaluate policy implications associated with these studies' and to "revisit" underlying technical and methodological issues in that context. For example, should studies include full assessments of health and safety risks to consumers and workers? How can the quality of data and findings be assured? What are appropriate and inappropriate uses of life-cycle information?

To help explore such issues in a changing policy landscape, the Steering Committee suggested that World Wildlife Fund & The Conservation Foundation convene a panel to offer diverse perspectives on the development and use of life-cycle studies. Twelve panelists, including several members of the Steering Committee, participated in a forum held on May 14, 1990, along with about 35 observers. Panelists represented consulting firms, major corporations, environmental organizations, academia, EPA, a metropolitan solid waste utility, a product labeling organization, and a government agency in the Netherlands.

The forum focused on issue identification rather than resolution of those issues. Specific objectives

---

*Various terminology has been used to denote studies relating to life-cycle characteristics including "cradle-to-grave" and "resource and environmental profiles." The term "product life assessment" was used for the forum described in appendix A of this report to convey the generic class of such studies for discussion purposes only. See following section for terminology adopted by the Steering Committee.

Figure 1.5
Selected Issues Identified at May 14 Forum*

DEFINING THE NATURE AND SCOPE OF PRODUCT LIFE ASSESSMENTS

- What *is* a product life assessment? What stages of the life cycle and types of analysis should be included?
- What are the reasons for conducting product life assessments? Should the scope of a study vary according to intended use of information? When, and how, could a "streamlined" study be conducted defensibly?
- Should risk assessments be addressed in studies? conservation of resources? economic analyses and valuation? product performance? aesthetic impacts?

POTENTIAL AND APPROPRIATE USES OF LIFE-CYCLE INFORMATION

- What are the various potential uses of life-cycle information? Which uses are appropriate and inappropriate?
- How should/will public decision makers apply life-cycle information? Can information be used as the basis for regulatory restrictions on products, materials, or substances?
- Can studies be used to generalize beyond specific products to address materials?

ANALYTIC METHODS AND DATA

- Should studies measure *amounts* of energy and materials used and wastes generated only or include associated *impacts*? Should, and how, could relative weights be assigned to various elements of the study, such as wastes generated within and across media?
- Is there a need for common methodologies? Should generic data bases be developed? How can the quality of information be assured? What are the constraints of confidential business information, and how can they be addressed?
- How should uncertainty be dealt with and presented? How far should studies go in terms of detail? When do expenditures outweigh benefits?

COMMUNICATING RESULTS OF LIFE-CYCLE ASSESSMENTS

- How can information best be communicated to the public? What are the tradeoffs between effectiveness in presentation and accuracy of information?
- Should all data and assumptions be made available for public inspection and what are the challenges and tradeoffs in doing so?

*Condensed from "Product-Life Assessments: Policy Issues and Implications—Summary of a Forum" (see appendix A.)

were to: 1) improve understanding of the potential contributions and limitations of life-cycle studies, 2) identify areas of the greatest uncertainty and potential conflict, and 3) provide background information for other efforts needed to resolve policy and technical issues. Appendix A provides more information on the forum including a summary of panelist presentations, discussion of various issues raised, and background materials distributed to forum participants.

A Life-Cycle Assessment Model

As figure 1.5 depicts, a wide range of issues were identified and discussed at the forum. Several of these issues reflected fundamental differences in participants' assumptions about the definition and purpose of a life-cycle study. Key examples were the debate concerning whether or not to include risk assessments in studies and whether or not pollutants should be evaluated in terms of impact or simply identified in terms of amounts. Differing views on such issues, of course, carried implications for associated methodologies, appropriate uses of information, and the communication of results.

Clearly, there is a need to develop a common understanding of study objectives if life-cycle information is to become an integral part of the evaluation process that results in source reduction improvements. The Steering Committee therefore recommends widespread adoption of the following model to facilitate that process.*

Collectively termed a "Life-Cycle Assessment," the model consists of:

1 an *inventory* of materials and energy used, and environmental releases (air, water, and land) from all stages in the life of a product or process;
2 an *analysis of potential environmental effects* related to the use of energy and material resources and environmental releases; and
3 an *analysis of the changes needed* to help bring about environmental improvements for the product or process under study.

By differentiating boundaries among the three stages, the model helps resolve many of the issues of scope and appropriate use raised at the forum. It clarifies that life-cycle studies conducted to date have focused on the initial inventory, which does not encompass risk assessment or other considerations of effects. Because of this more limited scope, inventories cannot be used as the sole basis for drawing conclusions about the relative environmental effects associated with alternative products or processes. All three parts of the model are needed to fully evaluate the policy implications of a product or process choice. While useful information can be gleaned from any single component of the model, most of the life-cycle information available today is restricted to the inventory stage. Considerable efforts are needed to develop the next two stages. Various applications and limitations are discussed below.

Part 1: The Inventory

As noted above, life-cycle studies performed to date can be characterized as inventories. As such, they provide useful information on the *quantities* or amounts of materials and energy used throughout the life cycle as well as environmental releases. They do not evaluate *effects* of those inputs and outputs and, therefore, should not be characterized or interpreted as measures of environmental impacts.

In addition to other purposes, information from the inventory can be used to:

- Compare certain energy, resource, and waste characteristics of alternative products and processes.
- Stimulate actions to improve resource efficiency and to reduce the amounts of wastes generated, where such actions do not require additional life-cycle analysis.
- Identify instances in which additional life-cycle analysis *is* needed to evaluate the relative effects of potential source reduction options. In such instances, the inventory provides baseline information for subsequent stages of analysis (that is, analysis of potential environmental effects and changes needed to bring about improvements).
- Target products, processes, and activities for more detailed evaluation. For example, a manufacturer might choose to evaluate source reduction options for a product or process that produces large amounts of wastes or that contains a substance known to be toxic.
- Help establish source reduction goals and to measure progress (see next section in this chapter on the need for setting goals).

The inventory is an important source of information for evaluating source reduction opportunities as detailed in chapter 2's Evaluation Framework (see Tool 3, part B, in particular.) Some activities do not always require additional analysis to determine whether or not they result in source reduction. For example, the lightweighting of aluminum cans over the years means that fewer resources are being used to make the product (for example, water, energy, materials) and fewer wastes are being released to the environment. Other examples include using products more efficiently (for example, using both sides of paper) or eliminating unnecessary packaging without corresponding alterations to the product inside. Such options generally do not involve the use of substitute materials or processes that would alter the types of resources used and wastes generated. Actions that eliminate a process or material might also produce straightforward reductions in wastes that would be considered source reduction (see list and description of source reduction options in Tool 2 of the Evaluation Framework.) In these cases—although there can be exceptions—the inventory generally does not so much determine *whether* source reduction occurs as it *documents* the types and amounts of reductions achieved.

While inventories do not include an evaluation of effects, they can help stimulate actions for doing so. The Toxics Release Inventory (TRI) collected pursuant to the Emergency Planning and Community Right-to-Know Act of 1986 is a good example of how inventoried amounts of releases alone have spurred ac-

*This model was developed at a workshop organized by the Society of Environmental Toxicology and Chemistry (SETAC) in August 1990. Workshop participants, including four Steering Committee members, met to define current methodology and to develop a technical framework for life-cycle studies.

tions at the federal, state, local and private levels to reduce wastes and to better understand the relative effects of various chemical releases. Likewise, many manufacturers have found that inventory-like tools, such as audits that track wastes from their sources to disposal, help identify for managers waste generation patterns that otherwise might not be detected or appreciated fully. Such awareness can lead to concrete actions to reduce wastes.

Inventories are by no means a simple tabulation of numbers. Research methodology is beset by challenges in getting the most reliable and complete data and typically requires a complex array of assumptions at each step of the life cycle—any one of which could potentially affect study results. Wholly different combinations of materials, energy, and wastes might be generated depending on the type of energy supply assumed (for example, hydropower, fossil fuels, nuclear power) and the location of those supplies.

Despite the complexity of analysis, a significant body of knowledge exists for carrying out inventories. And, while methodologies and assumptions will continue to vary according to client needs and consultant approaches, efforts are being undertaken to establish guidance on acceptable practices. The Society of Environmental Toxicology and Chemistry has outlined current methodologies and recommended procedures for the inventory stage of analysis.[35] In addition, EPA has initiated a two-year, multi-office project to: 1) help resolve outstanding issues associated with the scope and use of inventories (for example, data availability, data presentation, etc.) and 2) develop a framework for conducting analyses of potential environmental effects. EPA plans to draw on the expertise available outside the agency through peer review groups and advisors. These efforts can help standardize boundaries for the system under analysis, various factors to be measured, presentation of data, and other study features that could benefit from use of commonly accepted procedures. If widely adopted, such guidance can help address some of the concerns identified in the forum concerning quality control in use of the methodology (see appendix A.)

Part 2: Analysis of Potential Environmental Effects

The state of the art for the second stage of the Life-Cycle Assessment model is not nearly as advanced as for the inventory stage—largely due to its inherent complexity. This poses a significant challenge for source reduction analysis because some opportunities cannot be identified or fully understood without information on environmental effects. Knowing how much pollutant X is released in the course of making a product says nothing about the potential *effect* that pollutant may have in the environment. This can be an important consideration in making decisions about substituting one type of material or product for another, each of which may involve release of different pollutants with different effects. In short, information provided in the inventory often requires additional analysis to understand its significance.

Various types of analyses could be included in the second stage depending on the needs of the private- or public-sector decision maker. These could consist of both quantitative and qualitative measures of such effects as risks to public health, habitat modification, and aesthetic considerations such as litter and noise pollution. Interpretation of effects often will hinge on the user's priorities and concerns, in addition to factors such as geographic locale and boundaries of analysis over time and space. For example, interest in how much water is used in product manufacture likely will vary considerably depending on how scarce water supplies are in the affected region. Likewise, concern over the release of a toxic substance may vary depending on its toxicity, routes of release and exposure, concentration, and local habitat. Collecting and assessing data on these factors is a significant undertaking. Such complexities help explain why the prospects for devising standardized schemes for weighting environmental effects of pollutants often are met with skepticism (see appendix A for more discussion on weighting).

Do these challenges imply that analyses of effects cannot or should not be pursued? To the contrary. The emerging shift in policy emphasis from pollution control to source reduction largely reflects the growing recognition that existing laws and regulations, alone, *cannot* adequately address risks posed to public health and the environment at large. Pollution control programs were not designed to handle the types of complex and large-scale environmental problems identified today. Better information on the life-cycle effects of products, processes, and activities is fundamental to devising long-term strategies that get at the source of environmental problems. Better information also will be useful for federal and state appraisal of pollution control standards to determine whether they are set at appropriate levels. As demand for information on environmental effects continues to grow, so will efforts to devise tools that help supply that information.

In the context of municipal solid waste source reduction, information from the analysis of effects can be used to:

- make decision makers aware of environmental considerations that might be overlooked in the design, manufacture, and use of various products and materials—as well as of individual processes and activities of interest;
- select preferred alternatives among potential source reduction options that could not be selected without an analysis of effects;
- help establish source reduction goals and measure progress; and
- identify priorities for analyzing how to bring about source reduction improvements (in the third stage of the Life-Cycle Assessment model) or to trigger the need for a full inventory (in the first stage of the model).

As discussed previously, some source reduction options can be evaluated without information on effects, while others cannot. Substituting alternative substances or products may or may not reduce toxicity problems depending on the relative effects associated with those substitutes. Continuing with the aluminum can example, manufacturers generally believe that lightweighting of the aluminum beverage container may have reached its limits in terms of maintaining product performance. Any future steps likely will involve substitute materials and/or processes, which will require an analysis of effects to determine whether or not the changes result in source reduction.

In the case of household batteries, it is unclear whether substituting one rechargeable battery with about 20 percent cadmium content would reduce risks compared to the alternative of 200 to 300 disposable batteries that contain much smaller quantities of mercury (see appendix B). Which of these two particular alternatives is better in terms of source reduction? As a final example, what are the environmental effects associated with using disposable products versus alternatives that can be reused and/or recycled? Only an assessment of effects can provide the information needed to evaluate these types of trade-offs. (Such information could be used in Tool 3 of the Evaluation Framework to answer questions concerning life-cycle effects and trade-offs.)

Widespread attempts are being made to describe and measure effects and to develop indicators of potential effects. Geographic information systems and ecosystem simulation models are being developed and refined to model cumulative effects over time and space.[36] EPA has several projects under way to help advance methodology and application. It recently developed a two-volume screening guide to assist state and local officials and others in evaluating the potential risks posed by Toxic Release Inventory releases at particular locations.[37] EPA's Office of Toxic Substances is currently developing rules for a product stewardship program that asks manufacturers of toxic chemicals to develop information on the risks associated with the processing, use, and disposal of the chemicals they make. Finally, as noted earlier, several offices in EPA are cooperating in a multiyear effort to develop guidelines for Life-Cycle Assessments that can be used to gauge relative environmental attributes of alternative products and materials.

Part 3: Analysis of Improvements

The ultimate objective of life-cycle information, of course, is to bring about actual improvements in the quality of products, processes, and activities. The third stage of the model can draw from the inventory and analysis of effects to help identify needed improvements. It is a systematic evaluation, both quantitative and qualitative, of opportunities for using fewer raw materials or producing fewer releases that have adverse environmental effects. Manufacturers are most likely to conduct such analyses, but other sectors might also have the interest and resources for conducting certain types of studies in this category. Such improvements can involve both technical and behavioral source reduction options, identified and evaluated in Steps 2 and 3 of the Evaluation Framework, as well as the implementation strategies to execute those options, as discussed in Step 5.

Other Issues and Recommendations

Several issues raised at the forum tend to cut across all components of the Life-Cycle Assessment model. *Quality of data* is one of the most challenging. There are difficult trade-offs between using proprietary data and data that can be found in the public realm. Consultants often rely on confidentiality agreements to get access to site-specific and higher-quality industry data that would not otherwise be available. Lack of public access to such information, however, inhibits the public review and scrutiny many feel are needed to understand and verify results. For studies that are to be used in the public domain, data and methods should be available for review with full recognition that confidential or propriety information will need to be protected. In either case, data are not of the "lab bench" type that produce unequivocal results. Instead, studies often require various types of assumptions on a wide range of factors (for example, location of material inputs, regulations affecting production process, types of energy used, etc.) and careful interpretation. While concerns over data quality, proprietary needs, and public review are likely to remain, mechanisms such as use of peer reviews and the development of certain common data bases over time may help mitigate these trade-offs.

As discussed at the forum, there is considerable interest in developing *"streamlined" life-cycle assessments* or portions thereof. Conducting numerous inventories—let alone full life-cycle analyses—will require substantial investments in terms of time, human, and financial resources. In some applications, such as product labeling, there may be ways to limit the boundaries of analysis so that the inventory and analysis of effects is relatively narrow in scope (see chapter 3's discussion of product labeling). In many other cases, however, the Steering Committee discourages adoption of streamlined approaches in the absence of considerably more research. One of the merits of the Life-Cycle Assessment model is that it helps make decision makers aware of the various types of considerations that should be taken into account. Premature attempts to limit the range of considerations may be counterproductive in the long run and can lead to erroneous conclusions. Productive research might explore how to incorporate the full range of considerations in various levels of analytic detail and how to devise appropriate indicators of effects.*

Another topic of panel discussion concerned the issue of *weighting pollutants*, either within the media of their release (for example, air, water) or across media categories. Some of the previous inventory type of studies have summed quantities of pollutants to produce aggregate totals. While this has been done to simplify presentation of data, it implicitly assigns a one-to-one weighting to the pollutants and therefore makes assumptions concerning their effects. An itemized listing of all pollutants minimizes this problem and preserves the detail needed for subsequent analysis of effects.† This is the appropriate way to present inventory data, although data can be organized in various ways useful to readers such as appropriate regulatory categories. If weighting devices are attempted in an analysis of effects, the assumptions for doing so and their implications should be made explicit to the reader.

Communicating life-cycle information involves difficult trade-offs between simplicity and accuracy. If findings from an inventory, for example, are presented in too simple a fashion, the study loses accuracy; if presentation is too complex, a study can "die" of its own weight. Even if future studies are identified as being confined to an inventory scope and purpose, readers may misinterpret data to infer environmental effects associated with the products compared. The Steering Committee believes, however, that properly designed executive summaries can go a long way to adequately address these challenges. Full disclosure of major assumptions and data sources can improve public confidence in the study. Full discussion of what the findings mean, and don't mean, can safeguard against improper interpretation. Peer review mechanisms and focus groups can be used to help determine how to present information both simply and accurately.

Summary Recommendations

Among other recommendations and observations noted above, the Steering Committee recommends that:

- **The Life-Cycle Assessment model should be widely adopted by the professional and academic community involved in life-cycle analysis to help resolve many of the issues raised at the forum (described in appendix A).**
- **Follow-up workshops should be held to explore and make progress in implementing the remaining two stages of the Life-Cycle Assessment model.**
- **Inventories should not be used as the sole basis for drawing conclusions concerning the relative environmental effects associated with alternative products or materials.**
- **Inventory data on wastes should be presented as an itemized listing of their types and amounts to minimize problems of implicit weighting of environmental effects and to preserve the detail needed for successive analysis of effects.**
- **Peer review mechanisms and use of focus groups should be used in Life-Cycle Assessments, particularly for studies used in the public arena, to help ensure the quality of information produced as well as a check on the accuracy of its presentation.**

*EPA is currently investigating criteria for determining when various stages and levels of analysis under a Life-Cycle Assessment may be warranted.

†Other forms of implicit weighting can occur in the way in which data are collected and reported, such as the summing up of various organic compounds to report biological oxygen demand (BOD) in water.[38] See SETAC proceedings for discussion of weighting issues.

SETTING GOALS AND MEASURING PERFORMANCE

Municipal solid waste source reduction needs to be recognized—and implemented—as a national priority. Despite its great potential, source reduction policy is still in the early stages of development. Quantifiable goals would help stimulate and shape that development; measurement techniques would allow the tracking of its progress. Without this dual process, source reduction stands little chance of reaching its full potential.

Setting national goals is not, however, without significant challenges. Decision makers face an apparent dilemma between the need for accurate data and the need for action. Insufficient data and measurement tools exist upon which to base and track quantifiable goals; however, the very act of setting such goals can stimulate the types activities needed to generate sufficient data and measurement tools. This dilemma is compounded by the fact that it can be inherently difficult to measure waste that has not been produced.

Why Are Goals So Important?

Planning is fundamental to any successful and efficient enterprise. That process typically involves enunciating overall goals, setting objectives, ranking priorities, delineating specific steps and timetables, measuring performance, and factoring what has been learned into ongoing implementation plans and revised goals.

Goal setting is particularly crucial for source reduction. The magnitude of problems we face today in managing community wastes and conserving natural resources demands that source reduction opportunities be exploited to their fullest. That cannot occur without better understanding where we are going, and how much can be accomplished by various sectors and alternative strategies. Failure to have specific goals for source reduction exacerbates the tendency for people to confuse it with recycling, even though they involve completely different sets of behaviors and measurement issues.

Why Is Measurement So Difficult?

Measurement is the vehicle for tracking progress toward source reduction goals, evaluating policies for effectiveness, and revising plans when needed. Such a mechanism is particularly important during these early stages of policy development. As indicated above, measuring certain types of prevention activities can be relatively elusive compared with waste management techniques—that is, how can one measure what has not occurred? This apparent elusiveness, however, largely is a function of inexperience in measuring source reduction and a lack of readily available tools to document results. Many observers draw parallels with the state of measurement for recycling 10 years ago as it was beginning to generate more interest.

There are other challenges associated with measuring source reduction. First, policy makers need to determine appropriate benchmarks for measuring results. For example, considerable weight reduction has already taken place in some products, such as beverage cans. Should benchmarks reflect those previous efforts and the fact that further reductions may be negligible? Second, measurement tools need to be devised for comparing, and communicating, various types of life-cycle characteristics and effects. For example, how should changes in the volume of a product be evaluated against changes in the toxicity of wastes produced elsewhere in product manufacture?*

Finally, progress in source reduction needs to be evaluated within the context of other management priorities. For example, should a manufacturer pursue actions to make a product lighter or smaller if those changes inhibit recycling of the product or otherwise significantly affect product performance?

Steering Committee Recommendations

To address these types of challenges, the Steering Committee believes that goal setting for source reduction can best be accomplished by the development of two different types of goals, aimed at different purposes and different audiences. The first is a *national policy goal* that serves to inspire—rather than dictate—implementation plans by others. It should serve as the impetus for developing the second type of goal—*quantifiable goals for specific sectors of the economy and for individual organizations*. The following recommendations delineate needed actions and relative responsibilities for the various parties involved in source reduction at the national, sectoral, and individual organization levels.

National Goals

A national goal should clearly enunciate an overall policy direction to inspire continued progress toward actually reducing the wastes generated in the United States. Such a goal could serve as a catalyst for measuring and tracking source reduction and could

*See preceding section on Life-Cycle Assessments and appendix A for more discussion on this and related issues.

Figure 1.6
Examples of Source Reduction Goals

- **The U.S. Environmental Protection Agency, Municipal Solid Waste Program**: Reduce the rate of per capita waste generation and reduce waste toxicity.[1]
- **The National Governors' Association**: Reverse the growth in per capita waste generation and reduce the toxicity of consumer products. By holding per capita generation to 1985 levels (3.7 pounds per day), future waste stream volumes are estimated to be reduced by almost 16 percent overall.[2]
- **The Coalition of Northeast Governor's Source Reduction Council**: Their model legislation, already passed in at least eight states, will dramatically reduce the use and presence of four heavy metals—lead, mercury, cadmium and hexavalent chromium—from packaging manufactured or used in affected states.[3]
- **New York**: An 8 to 10 percent reduction by 1997 in the tonnage of the solid waste stream through source reduction. Seven states have source reduction goals—most of which include recycling.[4]
- **Procter & Gamble**: Each of the 100 or so brand-name products in the corporation has been charged with achieving at least a 25 percent reduction in waste generated before July 1992.[5]

help set priorities for research and funding. However, specific sectors of the economy face different technical and economic realities that make direct application of one overall goal to all sectors inappropriate. Also, complete data necessary for evaluating what is achievable over what period of time are incomplete. Therefore, a national goal should serve only as a general guide to specific implementation plans for individual sectors or organizations, at least at this time. The Steering Committee recommends that:

- **EPA should set a national policy goal for reducing the amount and toxicity of municipal solid waste based upon the best information currently available. Such a goal should set the overall policy direction for source reduction but should be seen as a target toward which to aspire, not as a uniform measure against which all sectors should be held accountable.**
- **EPA should also devise explicit mechanisms and measurement tools for generating the additional information needed to account for changes in the municipal waste stream attributable to source reduction, to report on progress toward source reduction goals, and to revise its national policy goal.**
- **EPA, state and local governments, industry, and consultants should cooperate to develop methods for studying and comparing municipal solid waste streams. Waste stream studies should be used to set baselines and allow measurement of source reduction.**

Sectoral Goals

Some sectors of the economy* may be able to achieve significantly *more* than the national goal; others may be able to accomplish *less* due to various constraints in both opportunities for source reduction and capabilities for their achievement. Further, goals generally imply some degree of accountability. This is most appropriately applied where sufficient data exist to establish confidence that a goal, although challenging, is reasonably achievable. Detailed, empirical data upon which to set feasible goals applicable to their waste streams is more easily available at the sectoral level than for the nation as a whole, but the difficulty in setting quantifiable goals is nevertheless extremely challenging. The Steering Committee therefore recommends that:

- **In consultation with the public and private sectors, EPA should conduct research and provide guidance on establishing quantifiable goals for the various sectors, as well as for developing measurement tools and tracking mechanisms.**
- **Where information is available, quantifiable goals should be developed for the various socioeconomic sectors of society (for example, a state or region, a commercial sector, or a major components of the waste stream, such as nondurable goods or yard wastes).**
- **The process of setting sectoral goals, like na-**

*Sectors may encompass both socioeconomic entities (that is, government, commercial, residential, and institutional) as well as major components of the waste stream (that is, durable goods, packaging, yard waste, etc.).

tional and organization goals, should be dynamic and progressive. As knowledge expands, and experience with source reduction measurement improves, sectoral goals should be refined and gradually made more specific. In addition, the aggregation of established sectoral goals should be used as the basis to periodically refine the national goal.

Organizational Goals

Empirical data upon which to set feasible goals are most complete and accessible for specific decision-making entities such as some state agencies and local governments, individual businesses, and trade associations. Actions at this level ultimately determine the degree to which source reduction is actually accomplished in the country. Even at this level, however, challenges in setting quantifiable goals and measuring progress exist. These challenges can be exacerbated for small communities and businesses that have fewer resources to collect and monitor necessary data. The Steering Committee recommends that:

- **State governments should set quantifiable goals for the institutional waste streams they produce, as well as for the waste streams within their jurisdictions, and should measure performance over time. Likewise, individual businesses should set and monitor source reduction goals for their products and operations as part of their overall pollution prevention plans.**
- **Individual organizations should consider other entities whose cooperation is necessary to set and achieve source reduction goals. For example, a supplier of goods or materials to another business interest should participate in the goal-setting process both as an individual organization and as a "player" in the overall production process that results in a given product and waste stream. Useful in its own right, this kind of cooperation may be particularly important for small organizations that have resource constraints.**

Chapter 2
Evaluating Opportunities for Source Reduction

How can the design and use of products be altered to reduce the amount and toxicity of municipal solid waste? Decision makers in both the private and public sectors must devise creative approaches for dealing with this question if current trends in the generation of wastes are to be reversed. Yet very little pragmatic guidance is available today to help identify and evaluate source reduction opportunities.

The Evaluation Framework presented in this chapter is intended to help provide that guidance. It consists of a five-step decision-making process analysts can use to evaluate the following basic questions:

- *Step 1 — What is the problem?*
- *Step 2 — Which source reduction activities (or "options") should be considered further?*
- *Step 3 — Based on a wide range of considerations, which options promise to be the most effective?*
- *Step 4 — Which implementation strategies might be considered to put the source reduction option into action?*
- *Step 5 — Which of those implementation strategies promise to be most effective?*

This chapter describes the Evaluation Framework, offers suggestions for its use, and presents the actual checklists and questions making up the framework. Examples of how the process might work are drawn from such products as household batteries, paint, third-class mail, small appliances, and food-service disposables (for example, single-service cutlery and dishware). In addition, an overview to the five-step process is illustrated using mail-order catalogues—both from the perspective of a state official working on solid waste and a management employee working for a mail-order company (see pp. 27-29). Appendix B contains a more detailed demonstration of the framework as applied to household batteries.

WHY IS AN EVALUATION FRAMEWORK NEEDED?

Some of today's most hotly debated issues about products and their use concern their relative effects on the environment. Which causes less of an impact, paper grocery bags or plastic? Cloth diapers or disposables? Paper plates or china? These are just a sample of the most commonly discussed questions faced by decision makers ranging from product designers and manufacturers to consumers and government officials. The underlying issue, however, is this: how can one tell if a change in the design and use of a product—or a change in a manufacturing process or activity—will actually result in less waste, or less toxic waste?*

Sometimes the answers to this question can be relatively straightforward. Chapter 3 provides examples of source reduction activities that can be undertaken today without the need for substantial new research. Often, however, answers can be extremely complex and hindered by various obstacles such as not having enough information to evaluate choices. Even where information is available, there are numerous other factors that are difficult to evaluate, such as possible trade-offs in product performance or price.

A more pervasive obstacle is the lack of guidance on *how* one should go about identifying and evaluating such questions in the first place. Where does one start? How can analysts systematically wind their way through the maze of possible choices and know they are asking the right questions along the way? The Evaluation Framework presented in this chapter addresses this obstacle.

---

*See chapter 1 for a discussion of Life-Cycle Assessments and how information from their various components may be used to determine whether or not an activity results in source reduction.

The Steering Committee recognizes that comprehensive analysis is needed to bring about meaningful reductions in the waste stream. To help decision-makers in public and private sectors evaluate source reduction opportunities, an Evaluation Framework is provided to guide and stimulate the creative approaches needed to get at the source of our waste generation problems.

WHO MIGHT USE THE FRAMEWORK?

The Evaluation Framework can be used by any type of decision maker interested in source reduction. Individuals involved in product manufacture, for example, likely will be familiar with the source reduction options most likely to apply to various products. Examination of the checklist used in Step 2, however, could stimulate ideas for applying other options to the product of concern in creative ways. A product designer might find the framework most useful as a checklist for the broad range of decision factors that might not normally be considered at the design stage. State municipal solid waste planners could use the framework as an overall guide to exploring opportunities and to identify questions that should be explored with designers or manufacturers of the product of concern. In short, the framework is best viewed as a source of ideas that can be drawn upon as needed by the individual user. It is hoped, also, that the framework serves as a bridge for expanding communication among the various parties that need to be involved in source reduction.

DESCRIPTION OF THE EVALUATION FRAMEWORK

The Evaluation Framework is a series of checklists and open-ended questions designed to stimulate thinking, highlight information needs, organize data, and help in making decisions. It is meant to encourage

Figure 2.1
Definitions Used in the Evaluation Framework

- ***Targets* for Source Reduction:** Components of the waste stream, product categories, and products are all terms that refer to potential targets for analysis. They form a hierarchy of increasingly specific categories. A *component* of the waste stream is a set of objects that behave similarly in the waste stream (for example, durable goods), a *product category* is a set of items with dissimilar functions but a common characteristic (for example, small appliances), and a *product* is a type of good within a product category that has a similar function (for example, blenders).* *For the purposes of this report, "product" shall refer to any of these three classes unless otherwise noted.*
- ***Manufacturers* and *Consumers*:** These terms are used here in the broad sense that manufacturers create a product or offer it for sale, while consumers use it. Manufacturing is not limited to the industrial sector; for instance, restaurants and retail stores can act as manufacturers. And consumers are not limited to individuals. For instance, both industry and government can act as consumers. A single entity can be both a manufacturer and a consumer. A company that fabricates television sets is both a manufacturer (of small appliances) and a consumer (of component parts). Manufacturing need not be limited to goods that are sold. A nonprofit group that mails out educational materials and requests for donations is a manufacturer of mail (as well as being a consumer of envelopes and stationery).
- ***Tools*:** Checklists of information or evaluation questions designed to assist decision making at each step in the evaluation framework.
- **Source Reduction *Options*:** Specific technical or behavioral actions for accomplishing source reduction in a product or its use. For example, manufacturers might be able to extend the life of a product by making it more durable. Consumers can purchase the more durable product, but can also extend its life through proper maintenance.
- **Source Reduction *Strategies*:** Policy actions that can be taken to implement or otherwise induce technical or behavioral options. For example, an awards program can encourage manufacturers to make a liquid product in a more concentrated form. Tax policies also can stimulate product concentration by making packaging more costly.

*Under this definition, packaging can also be analyzed as a "product."

both creative thinking to identify new solutions, and detailed analysis as the basis for making choices. Each step, with its accompanying "tools," is discussed in detail in the following pages. Four accompanying figures help to further illustrate these steps and tools:

- Figure 2.1 presents definitions of various terms used in proposing the Evaluation Framework.
- Figure 2.2 briefly summarizes each of the steps in the framework.
- Figure 2.3 illustrates how a hypothetical mail-order company might use the framework; and
- Figure 2.4 shows how a state regulatory agency might also use the framework to addresss the problem of waste generation from mail-order companies.

HOW TO USE THE EVALUATION FRAMEWORK

The Evaluation Framework is meant to be a source of ideas and to suggest factors that one may have overlooked. Tools are illustrative only and can be used or modified according to the user's interests, objectives and priorities. In short, the tools and five-step process are intended to stimulate creativity in identifying source reduction opportunities, not to prescribe a rigid formula for evaluation.

For example, users may start at different steps of the process. A state waste planner might wish to start at Step 1 in order to choose a product for analysis. A manufacturer, who already had a product in mind, could begin by looking at options in Step 2. Someone else who had already identified potential options could begin by analyzing those options in Step 3. Within an individual step, analysts may find that some questions are not always applicable. Finally, research may not always proceed in the linear sequence shown in the framework. Sometimes information on questions spanning several different tools may, in fact, be collected and evaluated simultaneously.

The tools for evaluating source reduction options and strategies are of two types: the checklists (Tools 2 and 4.B) allow one to quickly scan through and eliminate inappropriate options or strategies, while the evaluation tools (Tools 1, 3, 4.A, and 5) provide detailed sets of questions for analyzing specific options or strategies. Suggested matrices for Tools 1, 3, and 5 allow an analyst to qualitatively compare alternative options and strategies in order to select those with the most potential. Various "blank" forms for specific tools are included, to be copied and filled in by analysts wishing to use this approach.

AFTER THE EVALUATION: WHAT NEXT?

The ultimate purpose of evaluating source reduction opportunities, of course, is not to conduct analyses *per se* but to bring about actual improvements in products and their use. Effective presentation of evaluation findings can be critical in making that happen.

The five-step evaluation process described in this chapter can generate sizable amounts of information needed to help understand the targeted product, alternative options and strategies. It is suggested that analysts conclude their evaluation by preparing synopses of key findings, recommendations, and rationales, in order to pinpoint critical factors and to distill the information into meaningful decision-making documents. While the matrices of Tools 1, 3, and 5 are designed to facilitate that type of synopsis, and might be included in a synopsis, they are not substitutes for a well-reasoned discussion.

Summary documents are useful for purposes other than effective communication. The distillation process itself fosters critical appraisal by making key assumptions and methods more apparent. Summaries also can help streamline future evaluations by documenting such findings as which lines of inquiry proved more useful than others, etc. Of course, the advantages of making evaluations more efficient must always be weighed against the need to apply fresh objectivity and creativity in looking at different products.

Figure 2.2
Overview to Framework Steps and Tools

- **Step 1 — What is the problem?** The process begins by selecting a product (or "target") for analysis, acknowledging such considerations as whether to choose a product that causes major problems in the waste stream or one that is less of a problem but may be easier to solve.

  See Tool 1 (pp. 30-35) for seven types of criteria that might be used to select a product for analysis.

- **Step 2 — Which source reduction activities (or "options") should be considered further?** This step involves identifying the changes a manufacturer or consumer might undertake to reduce the problems associated with the selected product. Could it be made lighter or more durable? Could it be purchased in bulk sizes to reduce packaging? Imagination and common sense are required here, not detailed analysis.

  See Tool 2 (pp. 37-43) for a checklist with 30 different types of options, with space to tailor those options to the product selected. It is followed by a guide that explains what each option is and gives examples of its application.

- **Step 3 — Which source reduction option(s) promises to be the most effective?** Many factors need to be taken into account in trying to select the most promising source reduction options for a product. For example, what does the market look like for this particular product? What would be the spinoff effects for the business or other products if the source reduction option were to be implemented? This step requires the analyst to collect pertinent data, evaluate the options identified in Step 2, compare those alternatives, and choose which ones to recommend.

  See Tool 3 for collecting background information on the product itself (Part A, pp. 46-49) and for evaluating the source reduction options (Part B, pp. 49-52). Tool 3 includes a matrix for summarizing results of the analysis and for comparing options.

- **Step 4 — Which implementation strategies might be considered to put the source reduction options into action?** Once a potentially effective source reduction option is selected, steps must be taken to implement it. Step 4 involves identifying possible implementation strategies based, in part, on their ability to overcome obstacles to the source reduction options.

  See Tool 4.A (pp. 53-56) to help evaluate possible obstacles an option may face, such as technical difficulties or increased prices.

  See Tool 4.B (pp. 56-64) for a checklist of more than 60 different types of strategies—ranging from voluntary programs to incentives to regulations—followed by a brief description of how each might apply to source reduction options.

- **Step 5 — Which implementation strategies promise to be most effective?** Having completed the process, the user will now have a plan of action to initiate changes to the product under consideration.

  See Tool 5 (pp. 65-68) for questions concerning the effectiveness of the strategy, the feasibility of implementing it, and the burden on society. Tool 5 includes a matrix or summarizing resuts of the analysis and for comparing strategies.

Figure 2.3
A Mail-Order Company's Perspective

Theoretical example of how Company X (a mail-order company that wanted to reduce waste and save printing and mailing costs) might use the framework:

Step 1 — What is the problem? After reviewing screening criteria (Tool 1), company decided to analyze its catalogues, on the basis of criteria 7 ("Identified alternatives for source reduction") and 8 ("Other"—customer complaints about catalogues).

Step 2 — Which source reduction activities should be considered further? Company scanned checklist of options (Tool 2) and selected the following for analysis:

- *Option 2:* "Eliminate or reduce toxic substances in the product": Eliminate use of inks containing toxic substances;
- *Option 4:* "Lightweight/reduce volume": Make catalogue layout smaller, with lighter weight paper;
- *Option 8:* "Produce fewer models/styles": Send out catalogues quarterly rather than monthly (as is currently done).

Step 3 — Which options promise to be the most effective? Since Company X already had basic information on its own catalogues, it did not need to work through Tool 3.A ("Basic information for any option"), but next used Tool 3.B ("Information for evaluating a given option"). The company's preliminary analysis showed that consumers might not respond as well to catalogues with smaller pictures and type, or duller colors (due to non-toxic inks). In addition, without monthly catalogues to remind consumers of company's products, customers might order from competitors. The company thought that these losses could be offset by monetary savings from using less paper and reducing mailing costs. In addition, the company hoped to capitalize on a "green marketing" strategy to appeal to current customers and perhaps to attract new customers. Company decided to proceed with analysis of all options selected in Step 2.

(Figure 2.3 continued on next page.)

Figure 2.4
A State Government Agency's Perspective

Theoretical example of how a state government official working on solid waste might use the framework to address the same problem (waste generation from mail-order companies):

Step 1 — What is the problem? Official selected category of catalogues after reviewing Tool 1 screening criteria because of criteria 1 ("Percentage share of the waste stream"), 2 ("Expected growth in quantity and share of MSW"), and 8 ("Other"—public concern).

Step 2 — Which source reduction activities should be considered further? Official scanned checklist of options (Tool 2) and selected the following for analysis:

- *Option 1:* "Eliminate product/reduce amount": Convince companies not to send catalogues to those who don't want them and encourage consumers to take their names off unwanted mailing lists.
- *Option 2:* "Eliminate or reduce toxic substances in the product": Convince companies to reduce/eliminate use of inks containing toxics.
- *Option 4:* "Lightweight/reduce volume": Convince companies to make catalogues smaller, with fewer pages; and using a lighter basis weight paper.
- *Option 8:* "Produce fewer models/styles": Convince companies to send out catalogues less often.

Step 3 — Which options promise to be the most effective? Official gathered information to answer questions in Tool 3.A, evaluated options using Tool 3.B, and decided to proceed with analysis of all options selected in Step 2. Official noted that pursuing Option #1 through a regulatory strategy could force companies to advertise in other media (like newspaper inserts) that might not reduce the quantity of waste, and could actually increase it.

(Figure 2.4 continued on next page.)

Figure 2.3 (cont.)
A Mail-Order Company's Perspective

Step 4 — Which implementation strategies might be considered to put the source reduction option into action?

Tool 4.A — Company identified obstacles for each option selected in Step 2:

- *Option 2:*
  — *Technical obstacle:* More work needs to be done to develop nontoxic inks in as broad a range of colors as inks containing toxics currently used;
  — *Consumer preference/economic obstacle:* If consumers react negatively to catalogues printed with current generation of non-toxic inks (less attractive), sales could drop;
  — *Economic obstacle:* Nontoxic inks may cost more.
- *Option 4:*
  — *Consumer preference/economic obstacle:* Smaller catalogues might be hard to read and pictures might be unclear, causing a drop in sales; consumers may react negatively to lighter weight paper.
- *Option 8:*
  — *Consumer preference/economic obstacle:* With less frequent catalogues, consumers might forget about company and purchase from another source.

Tool 4.B — Company scanned checklist of source reduction strategies for each option:

- *Option 2:* Company identified two strategies:
  — *Strategy I.J ("Stimulate consumer demand for alternative products"):* Provide information to customers on the appearance and other characteristics of nontoxic inks to mitigate a drop in sales if a switch were made.
  — *Strategy I.M ("Corporate research"):* Encourage research to develop nontoxic inks that would be as attractive and affordable as inks currently being used.

Figure 2.4 (cont.)
A State Government Agency's Perspective

Step 4 — Which implementation strategies might be considered to put the source reduction option into action?

Tool 4.A — Official identified obstacles for each option selected in Step 2:

- *Option 1:*
  — *Information obstacle:* The public needs information on how to stop receiving unwanted catalogues.
  — *Consumer preference/economic obstacle:* Sending fewer catalogues might reduce sales.
- *Option 2:*
  — *Technical obstacles:* More work needs to be done to develop attractive, less-toxic inks.
  — *Information obstacle:* Companies may not be aware of the availability of non-toxic inks.
- *Option 4:*
  — *Consumer preference/economic obstacle:* Smaller catalogues might reduce sales.
- *Option 8:*
  — *Consumer preference/economic obstacle:* Less frequent catalogue might reduce sales.

Tool 4.B — Official scanned checklist of source reduction strategies and chose the following:

- *Option 1:*
  — *Strategy I.D ("Provide services that promote source reduction activities"):* Promote Direct Marketing Association's Mail Preference Service that allows consumers to request that their names not be rented to other companies.
  — *Strategy I.S ("Write letters or take other action to influence government and producers"):* Conduct media campaign encouraging public to contact mail-order companies asking them to purge their mailing lists and make it easy for people to get off the list (such as providing a pre-paid card that cus-

- *Option 4:* Company identified two strategies:
  - — *Strategy I.J ("Stimulate consumer demand for alternative products"):* Conduct media/public outreach campaigns to change preferences of some customers for a large catalogue; capitalize on "green marketing" strategies.
  - — *Strategy I.N ("Set voluntary goals for source reduction"):* Identify a source reduction goal for the company to work towards in terms of a smaller catalog.
- *Option 8:* Company found that strategies I.J and I.M could also be applied to this option.

Step 5 — Which of these implementation strategies promise to be most effective? Company used evaluation questions for selecting strategies (Tool 5) and reached the decision to:

- Conduct further research into procuring environmentally benign inks that would be as attractive and affordable as inks currently being used; contact industry association for assistance.
- Conduct pilot project in which the company would send out slightly smaller catalogues printed with nontoxic inks to some customers and evaluate the response. Another pilot would test sending catalogues to some customers on a quarterly basis, with occasional small fliers to advertise "sale" items.
- Engage in "green" advertising (within new catalogue, existing advertising sources, and new advertising markets) to promote consumer awareness of new environmental stance.
- If pilot projects prove successful, expand to entire customer list.

tomers could return to remove name from list).

- *Option 2:*
  - — *Strategy I.F ("Government dissemination of data on source reduction"):* Provide information on suppliers of nontoxic inks and the performance characteristics of different inks.
  - — *Strategy III.17 ("Ban on products"):* Ban toxic inks in catalogues.
- *Option 4:*
  - — *Strategy I.S ("Write letters..."):* Conduct media campaign encouraging public to contact companies and request smaller catalogues, and to patronize companies that do so.
- *Option 8:*
  - — *Strategy I.S ("Write letters..."):* Conduct media campaign encouraging public to request that catalogues be sent on a less frequent basis, and to patronize companies that do so.
  - — *Strategy II.12 ("Tax on products other"):* Impose a waste tax on companies mailing out catalogues more often than quarterly.

Step 5 — Which of these implementation strategies promise to be most effective? Official used evaluation questions for selecting strategies (Tool 5) and reached the decision to:

- Proceed with both strategies under option #1 (promote Direct Marketing Association service and encourage public to request that manufacturers remove their names from mailing lists).
- Proceed with strategy under option #2 to disseminate information on nontoxic inks but abandon strategy to ban toxic inks due lack of resources to overcome obstacles in short term.
- Proceed with strategy I.S under options #4 and #8 (conduct a media campaign encouraging public to contact companies requesting smaller catalogues, lighter weight paper, and less frequent catalogs).
- Abandon strategy II.12 under option #8 (proposal to tax companies sending out catalogues frequently) due to questionable legal authority and jurisdictional constraints.

Step 1
Select Target in the Waste Stream

(Use Tool 1 — Screening Criteria for Selecting Priorities)

Step 1 of the evaluation framework requires analysts to select "targets"* as priorities for analysis. Tool 1, presented below, is provided to assist in that step.

Selecting potential targets for source reduction requires basic information on products of concern in the municipal waste stream that are of interest to the specific user. For example, if the analyst is most interested in reducing the *quantity* of waste needing municipal management, he or she will need information on products that contribute the most volume to landfills in the region of concern. If the user is more concerned with *toxicity* problems, the information needs are quite different. If an analyst already has a product in mind, he or she can skip Step 1 and start with Step 2.

Screening Criteria

Tool 1 describes eight different criteria that can be used to select potential targets for source reduction. These criteria are not meant to be all-inclusive, and users are encouraged to modify them to suit their own particular needs. In summary, the suggested criteria are:

1. Percentage share of the waste stream,
2. Expected growth in quantity and share of MSW,
3. Toxicity,
4. Wastes generated over the product life cycle,
5. Special handling considerations,
6. Availability of information,
7. Identified alternatives for source reduction, and
8. Other considerations.

Explanations of these criteria are provided below, with discussion of the types and sources of information that could be collected. A summary matrix for selecting targets is included as a tool on page 35. Please feel free to copy (or modify) it for your own use. Following these explanations, a ranking system is proposed to simplify the process of choosing products for further study. While this system is intended to be helpful, it stops short of prescribing how to make "apples and oranges" comparisons, such as whether volume or toxicity is more important. That decision is left to the user.

Tool 1
Screening Criteria for Selecting Priorities

1. Percentage Share of the Waste Stream

Since diminishing landfill capacity is one impetus for source reduction, this criterion measures the product's contribution to the volume of municipal solid waste (MSW). Two types of approaches have been used to estimate the amounts of materials and products generated and/or discarded.† One approach is based on actual measurements or surveys of MSW composition. The other approach, a material-flows methodology, converts production data to estimates of materials and products generated, recovered, and discarded.

The site-specific approach, typically used by local and state entities responsible for managing local waste streams, involves physical sampling of materials discarded. Two such examples are a study by the Rhode Island Solid Waste Management Corporation, and one performed by the Matrix Management Group for the Seattle Solid Waste Utility. While variations in local conditions, definitions, and measurement techniques inhibit their usefulness for comparative interests, they can be vital sources of information for local officials.

*This report uses the term *product* to describe either components, product categories, or products. See earlier discussion of "Targets" on page 24.

†The estimates discussed here represent only the disposal of the final product—not the wates generated from manufacturing or raw material processing.

The Garbage Project of the University of Arizona's Department of Anthropology has also conducted various analyses of MSW composition based on actual measurements. Their research on landfill excavation and detailed community profiles has contributed substantial new information on waste generation.

The only nationwide sources of information on the percentage share of various products in the waste stream are the periodic analyses prepared for the U.S. Environmental Protection Agency (EPA) by Franklin Associates, Ltd. EPA's analyses use the materials flow methodology. The 1990 update provides considerably more information for analysts than earlier efforts, including new emphasis on MSW generation rather than discards after recovery, supplementary estimates on volume rather than weight alone, some comparison with site-specific results, and additional detail on such components as plastics, lead-acid batteries, and diapers.[1]

Other important sources of information include the Office of Technology Assessment's 1989 report on solid waste and EPA's 1989 Agenda for Action report, with its various detailed appendixes.[2] OTA's report, for example, evaluated sample information on MSW generation rates from 28 cities and 9 counties as a tool to help understand trends and data deficiencies.

Most of the difficulties associated with estimating MSW stem from inconsistent use of definitions and measurement techniques among the numerous waste composition studies conducted. While there always will be the need to tailor studies for local needs, improvements in standardized data collection and reporting will be necessary for comparative purposes (as well as to improve our understanding of generation and disposal trends, to set goals, and to track progress).

The data from these sources often are aggregated into broad classifications like "major appliances" and may not provide the level of detail necessary to analyze specific products. In these cases, it will be necessary to gather information on sales volume, which can act as an indicator for the eventual volume of waste to be disposed. Sources of such information include associations, trade journals, and manufacturers of the product.

2. Expected Growth in Quantity and Share of MSW

The previous criterion (percentage share of the waste stream) measures the amount of a product in the waste stream now or at some date in the past. But information on future quantity (in the absence of any source reduction activities) is important in choosing among alternatives for analysis. This criterion is based on predictions of future volume. Quantity refers to the size of the waste stream for this particular product, measured in terms of volume or weight. Share refers to the fraction of the total waste stream comprised of this product. (Since the total volume of MSW is increasing so rapidly, a product's share of the waste stream can decrease even if its volume is increasing.)

The analysis performed by Franklin Associates for EPA also contains estimates for growth. Again, it may be necessary to consult associations or trade journals for projections of sales figures for specific products, which can act as surrogates for the quantity eventually disposed of.

3. Toxicity

Comprehensive information on toxic substances in MSW—including the distribution of substances in various products, what exposures are sufficient to cause problems, and what relative risks those substances pose under different MSW management systems—will vary depending on the substance under consideration. For instance, there is relatively extensive data on lead and cadmium,[3] but information on many organics is unlikely to be available. Simple numerical measures do not exist for toxicity, as they do for volume and expected growth. Instead, this section should contain a qualitative description of the toxicity concerns based on the limited information available.

4. Wastes Generated over the Product Life Cycle

As noted in the definition of source reduction in Chapter 1, source reduction should not increase the net amount or toxicity of wastes generated throughout the life of a product. This criterion accounts for potential environmental impacts arising over a product's entire life. Impacts include amounts of natural resource consumption and waste released to air, land, and water from extraction, production, use, and disposal, as well as the effects of these activities. (See chapter 1 and appendix A for discussion of issues in life-cycle assessments.)

Again, there are no easy numerical answers. Few, if any, products have had full life-cycle assessments performed, so it is unlikely that comprehensive data are available to identify the wastes generated and their environmental effects. If an inventory alone is available (that is, one that identifies the types and amounts of wastes), that may be sufficient to identify targets for analysis based on the presence of large amounts of wastes or potentially toxic substances. Such targets may or may not require additional life-cycle analysis of effects. In the absence of any life-cycle information, all that can be done is to include a qualitative discussion of the resource consumption and pollution impacts that are of particular concern for this product. A more complete evaluation may be performed if the product is selected for further analysis. Products that have significant negative impacts outside the municipal solid waste stream might be given priority when screening candidates for source reduction analysis.

5. Special Handling Considerations

Some segments of the waste stream demand special handling because their disposal could otherwise cause problems due to toxicity (such as waste oil or household hazardous waste) or other characteristics (such as the tendency for tires to float to the surface of a landfill). This criterion accounts for such characteristics. It does not cover items that may be collected separately but do not present special handling problems (such as newspapers collected for recycling, or furniture, which may be collected separately because of its bulk). Some users may want to target such products for source reduction, with the rationale that this would eliminate the need for special handling and reduce the environmental impacts of improper disposal (since some fraction is always disposed of outside the collection program). On the other hand, other users may assume that such products are adequately managed in special collection programs. Instead, they will want to target products that do not have special collection programs (because they are not concentrated enough to collect or are not economical to recycle), on the basis that source reduction may be the only way to remove such products from the waste stream.

6. Availability of Information

A great deal of data will be necessary to analyze a product. Lack of information will be a significant constraint in evaluating source reduction opportunities. Thus, the availability of information should be an explicit consideration when choosing products. The quality of the information should also be considered. Users can have more confidence in material that has been verified or peer-reviewed.

Again, relative availability of information can be a criterion arguing for or against selection. It will be easier to analyze a product if information is readily available, and more likely that specific recommendations for source reduction can be made. Alternatively, it is important to gather data for products on which there is scant information, in order to further develop the basis for future analysis.

7. Identified Alternatives for Source Reduction

Existence of feasible options might help differentiate cases where recommendations will be relatively easy or difficult to develop and/or implement. (Feasibility is related to such factors as the cost and performance of the alternatives.) Automotive oil and lead-acid car batteries are two examples of products that currently lack feasible alternatives.

Although the analyst has not gone through the detailed information gathering process of Tool 3, enough information usually exists to have a common sense idea of whether alternatives exist and how feasible they are.

8. Other Criteria

Among other possible considerations are at least three that do not fit under any of the above criteria. First, *the social good or need served by the product* is critical because some products, like tamper-resistant packaging, may provide such benefits to society that they may not be candidates for source reduction even when alternatives exist. Second, *public concern* about the product and the problems it creates may indicate a willingness on the part of the public to make changes that are necessary to achieve source reduction. Alternatively, a source reduction analysis can increase public awareness and concern. Third, *industry's degree of concentration* (the number of firms that manufacture the item and their market share), may indicate how successful one can be in implementing certain source reduction alternatives, since the degree of concentration may affect how easily manufacturers can be influenced and what results a change by a single manufacturer will have in the marketplace.

This criterion, in particular, is one that users may wish to modify depending on their interests.

Use of the Criteria in a Matrix

One way to select products for further analysis using these criteria would be using a ranking system and a matrix. This would involve developing a series of rankings for each criterion and then assigning a rank to each product being considered for analysis. This information could then be displayed in a matrix. The criteria would be listed along one axis and the products along the others, with the rankings filling the matrix in (see example on p. 34).

The benefits of such a matrix are that it simplifies the information under each criterion (for example, into low, medium, and high) and that it makes it possible to "visually assess" all of the information on one page. *It is unlikely that weights for the rankings, which would allow an overall comparison between products based on a single score, could be agreed to.* Different users will have different objectives, and thus different priorities. A government agency with responsibility for landfills may be most concerned with the criteria for share and growth, while an official in charge of composting programs may be most concerned with toxicity. A manufacturer, however, might give the highest weight to alternatives and life-cycle waste. Users of such a matrix will have to decide which products to choose based on their own subjective evaluation of the trade-offs between different criteria.

The figures that follow present one possible system to describe ranking options for these criteria (2.5), a matrix showing how the rankings could be applied to a set of hypothetical products (for illustrative purposes, 2.6), and a blank matrix for the user to fill in (2.7). The generic products are not meant to represent any particular products. It is not indicated which of these products would be chosen for analysis, because this decision is a matter of the user's judgment.

Figure 2.5
Summary Evaluation of Product Categories

Category	**Possible Ranking System**		
1. Share	*	=	Comprises less than 1% of MSW (specify whether this is on the basis of volume or weight).
	**	=	Between 1% and 5% of MSW.
	***	=	Over 5% of MSW.
2. Growth	*	=	Share of MSW to decrease over the next decade, although weight or volume may increase.
	**	=	Share of MSW to increase up to 10% over the next decade. (This refers to the rate of increase of the share, not the absolute change in share.)
	***	=	Share of MSW to increase more than 10% over the next decade.
3. Toxicity	*	=	Product contains low level of toxics or is nontoxic (e.g., tissue paper).
	**	=	Product contains some toxic components (e.g., a plastic containing heavy metal additives for coloration).
	***	=	Product contains high levels of toxics (e.g., household hazardous waste)
4. Life-Cycle			No ranking system developed because this is a multi-attribute function. Ranking will depend on what attributes of alternatives are considered: energy use, air pollution, water emissions, etc. It may be necessary to develop subcategories, with a ranking for each sub-category. Or this may be a category where no simplified ranking can be made.
5. Special Handling	*	=	Does not cause handling problems.
	**	=	Disposal may cause problems but does not need to be handled separately from other waste.
	***	=	A problem waste that may need to be handled separately from other wastes.
6. Information	*	=	Relatively complete information is available.
	**	=	Some information is available.
	***	=	Little to no information is available.
7. Alternatives			No ranking system was developed, because this is a multi-attribute function.
8. Other	x	=	Flags the existence of other considerations (no ranking).

Figure 2.6
Sample Summary Matrix for Selecting Targets
(products are hypothetical)

	Criteria							
	Contribution		**Environmental Impacts**		**Other Considerations**			
Category	% Share	Rate of Growth	Toxicity	Life-Cycle Impacts	Special Handling	Available Information	Feasible Alternatives	Other
Durables								
Product A	**	*	**		***	**		x
Product B	***	**	**		**	*		
Product C	**	**	?		*	*		
Product D	**	*	*		***	**		x
Product E	*	*	***		***	**		x
Nondurables								
Product F	***	**	*		*	***		x
Product G	**	***	*		*	**		
Product H	**	*	?		*	*		
Product I	*	?	***		**	**		x
Product J	**	*	***		**	**		x
Product K	?	?	*		*	*		
Disposables								
Product L	**	*	*		**	**		

Ranking System: *** = High, ** = Medium, * = Low, NA = Not Applicable, ? = Not known, x = See detailed description.

Figure 2.7
Summary Matrix for Selecting Targets

Criteria

Category	Contribution		Environmental Impacts		Other Considerations			
	% Share	Rate of Growth	Toxicity	Life-Cycle Impacts	Special Handling	Available Information	Feasible Alternatives	Other
1.								
2.								
3.								
4.								
5.								
6.								
7.								
8.								
9.								
10.								
11.								
12.								
13.								
14.								
15.								
16.								

Possible Ranking System: *** = High, ** = Medium, * = Low, NA = Not Applicable, ? = Not known, x = See detailed description.

Step 2
Scan the Range of Source Reduction Options

(Use Tool 2 — Checklist of Source Reduction Options)

Once potential targets of source reduction have been selected, it is time to scan the range of reduction options that are available. Tool 2, the "Checklist of Source Reduction Options," is designed to help determine the range of technical and behavioral alternatives available in any particular situation to accomplish source reduction. This checklist is composed of source reduction options for both manufacturers* (who can alter their products) and consumers (who can alter their purchases or habits). This checklist appears on page 38. Please feel free to copy (or modify) it for your own use. After a brief explanation of how to use this tool, the text below describes how each option could accomplish source reduction (to clarify the checklist); also included are real-life examples (to stimulate ideas on how the option might apply to a particular product).

How to Use Tool 2

The checklist in Tool 2 should be used by checking as many options for source reduction as common sense indicates might be relevant and then describing in the space beneath how the option could apply specifically to the product under consideration. For example, if the product is third-class mail, one option will be "Produce fewer models/styles" (Option 8 on the checklist), which in this particular case might be accomplished by sending out catalogs quarterly, instead of once a month.

In many cases, particular action may fall under more than one heading. For instance, one option for reducing the volume of commercial mail is for mailers to eliminate separate return envelopes, by using envelopes with a flap that allows recipients to turn the original envelope into a payment envelope. This can fall under either the "Combine functions of more than one product," "Produce for consumer reuse," or "Improve efficiency" option. Remember: the checklist is designed to elicit information and stimulate thinking. What is important is that the analyst identify "eliminating return envelopes" as an option, and not worry about what heading it is to be placed under.

There may be be more than one tailored option under a specific heading. With household batteries, for example, "Purchase improved complementary products" might include purchasing more efficient battery-operated devices (toys, radios, etc.) and purchasing more sophisticated battery chargers for rechargeable batteries.

Tool 2 can be used with either of two approaches. The first—and the one recommended for most analysts—is to eliminate obviously inappropriate options (or conversely to check off all the options that might possibly apply to the product selected). At this step it is better to err on the side of *inclusion* rather than exclusion. This approach encourages the sort of preliminary "brain-storming" that is useful prior to moving on to Step 3 (actual selection of options) and minimizes the risk of overlooking options with possible merit.

The second approach, which may work when one wishes to focus on one or more options either to test the process or because of a unique opportunity, is to begin analysis with only those options that seem to be most applicable. This approach minimizes the amount of analysis needed early in the process and assumes that the most obvious options are likely to be among the most promising. This streamlined approach does not prevent an analyst from adding other options during the evaluation process, but it does not facilitate that process. *Analysts may wish to experiment with both approaches to see which works best for their own circumstances.*

Understanding the Various Options

Not surprisingly, no one option will result in an environmental improvement in every application. Thus, the appropriateness of the various options described below must be considered on a case-by-case basis. Sometimes, selection of one option at one point in the

*As noted in figure 2.1 (page 24), the terms *manufacturer* and *consumer* are used here in a broad sense.

manufacturing process can result in increased waste elsewhere in the process—for example, switching to a lighter weight material could cause an increase in pollutants elsewhere in the life cycle. Nor will consumers accept all of these changes for all products. Inclusion on the checklist is not intended as a recommendation to apply an option in all cases. Step 2 is designed only to scan the list of options. Evaluating the desirability of the options does not occur until Step 3.

Note also that the examples below are intended to illustrate how an option might be applied to different products or product users. No attempt was made to verify whether the examples listed below actually resulted in source reduction. Mention of a product or manufacturer does not represent an endorsement by the Steering Committee.

Tool 2
Checklist of Source Reduction Options

Tool 2 is illustrated in figure 2.8. In producing this checklist, the Steering Committee used the following definitions.

1. Eliminate Product or Reduce Amount / Don't Purchase Product or Reduce Use

Manufacturers could stop producing, and/or consumers could stop purchasing, a particular product of concern. This option generally assumes the product is not essential.

- Certain packaging can be eliminated without harming product contents or their safe handling. For example, outer boxes on some brands of shampoo and deodorant have been eliminated.
- Consumers can avoid impulse purchases of unnecessary goods.
- Consumers can ask to be removed from mailing lists.
- McDonald's reports that eliminating the dividers from the shipping cases in which it receives cold cups reduced the weight of each case by eight ounces, which adds up to two million pounds of material saved a year.
- Consumers can use a plumber's snake to open clogged pipes instead of corrosive drain cleaner.
- Homeowners can avoid costly sprinkler systems by planting native species of plants or drought-tolerant plants that grow with the natural rainfall of the area and by mulching these plantings with yard wastes such as grass clippings, leaves, and chipped prunings. Homeowners can also reduce or eliminate the use of fertilizers.

2. Eliminate or Reduce Toxic Substances in the Product / Purchase Product with Reduced Toxics

Manufacturers could eliminate or reduce the amount of toxic substances contained in the product or the process to produce it without causing a net increase in the toxicity of wastes from its production. Consumers could facilitate this option by purchasing improved or preferable products.

This option may involve the substitution of nontoxic or less toxic inputs or products, as in the substitution of naturally based pesticides for petrochemically based ones, or nontoxic latex paint for oil-based paints. Or it may involve a simple reduction in the quantity used, without substitution. An example is the large decrease in the amount of mercury used in alkaline batteries, achieved by increasing the purity of other inputs, and improving manufacturing techniques.

- Two German companies, Livos and Auro, manufacture paints and a variety of other wood-finishing products (stain, varnish, shellac, wood preservative, thinners, etc.) without synthetic substances. Instead, they use materials such as linseed oil, pine resin, and citrus oil.
- Safer, Inc,. produces pesticides based on fatty acids that are targeted toward specific pests.
- Homeowners can use Integrated Pest Management (IPM) strategies instead of purchasing pesticides. Examples of IPM strategies include:
  - — Controlling rodents by preventing their access to food and nesting sites inside the house; controlling roaches by using boric acid instead of pesticides.
  - — Using plantings in the garden that attract parasites and predators of pest species. For example, allowing plants in the Umbellifer Family (such as carrots, fennel, and celery) to flower in the garden attracts parasites and predators of aphids.
- Red garbage bags are used to segregate infectious waste from ordinary solid waste in hospitals, doctor's offices, labs, etc. After discovering that

(continued on p. 40)

Figure 2.8
Tool 2 — Checklist of Source Reduction Options

	Manufacturers	Consumers
1.	__ Eliminate product or reduce amount	__ Don't purchase product or reduce use
2.	__ Eliminate or reduce toxic substances in the product	__ Purchase product with reduced toxics
3.	__ Substitute environmentally preferred materials or processes	__ Purchase environmentally preferred products
4.	__ "Lightweight" or reduce volume	__ Purchase reduced products
5.	__ Produce concentrated product	__ Purchase concentrated products
6.	__ Produce in bulk or in larger sizes	__ Purchase in bulk or in larger sizes
7.	__ Combine functions of more than one product	__ Buy multiple-use products
8.	__ Produce fewer models or styles	__ Purchase fewer models or don't replace for style
9.	__ Increase product life span	__ Purchase long-lived products

10. __	Improve repairability	__ Maintain properly / repair instead of replace
11. __	Produce for consumer reuse	__ Purchase reusable product / reuse product / donate to charity
12. __	Produce more efficient product	__ Purchase more efficient product
13.	N/A	__ Use product more efficiently
14. __	Change characteristics of complementary products	__ Purchase preferred complementary products
15. __	Remanufacture product	__ Purchase remanufactured product
16.	N/A	__ Borrow, share, or rent product
17. __	Other	__ Other

some companies manufacture red bags that contain enough lead to render fly ash hazardous for a typical incinerator, Waste Management, Inc., Medical Services convinced their suppliers to eliminate use of heavy metals in pigments and inks.

3. Substitute Environmentally Preferred Materials or Processes / Purchase Environmentally Preferred Products

It may be possible to switch to inputs or processes that are less damaging to the environment in terms of energy and materials use. Switching from chlorofluorocarbons (CFCs) to hydrochlorofluorocarbons (HCFCs), which are less destructive to the ozone layer, is one example. Such changes do not necessarily affect MSW significantly. They are included because municipal solid waste source reduction acts to conserve resources as well as reduce pollution, and extends beyond the solid waste stream to the entire life cycle of products and materials.

- Vitronics, a new company in New Hampshire, uses a natural terpene, derived from orange peels, as a substitute for a CFC-based solvent to clean electronic circuit boards.
- Chlorine-based bleaching of paper releases numerous chlorine compounds in the plant effluent and retains some level of them in the pulp. Associated environmental impacts of paper production can be reduced by using paper with lower brightness levels (so it requires less bleaching), using chlorine substitutes such as hydrogen peroxide, or using unbleached paper.
- Consumers can replace caustic oven cleaner with noncaustic products (such as liquid soap and borax, or baking soda and salt) and "elbow grease."

4. "Lightweight" or Reduce Volume / Purchase Reduced Products

A manufacturer could use these options to reduce either the weight or size of a product without changing its function. "Lightweighting" refers to any process that makes a product thinner and/or lighter without compromising its structural integrity, as has been done for beverage containers and newspaper. "Reduce" refers to making the product smaller, as has been achieved with many appliances and other electronic devices through miniaturization.

- Since the mid-1960s, the weight of—and hence material used in—the aluminum beverage can has been reduced about 35 percent.
- McDonald's was able to make its straws 20 percent lighter after switching to a different material, eliminating one million pounds a year.
- Redesigning its plastic bottles allowed Crisco Oil to lightweight by 25 percent.
- Consumers and retailers can ask suppliers to reduce packaging.

5. Produce / Purchase Concentrated Product

This option describes making a product less dilute by removing water, filler, or impurities to reduce the space it occupies. Because they generally require less packaging and less energy to transport, concentrates conserve resources and often save money.

- Downy Refill (a liquid fabric softener) is a concentrated product designed to be mixed with water in a used plastic Downy bottle. The refill package is 75 percent smaller than the diluted alternative.
- Procter & Gamble's Cheer concentrate uses 50 percent less packaging per washer load than its predecessor.
- By switching from ready-to-serve orange juice to concentrates, McDonald's was able to reduce packaging use by about four million pounds a year.

6. Produce / Purchase in Bulk or in Larger Sizes

Producing a product in bulk, or in larger sizes, serves to eliminate or reduce the amount of packaging needed to deliver a product to the consumer. Bulk goods, such as dry goods sold in supermarkets, are not prepackaged but are measured and sold in the quantity desired. Take-home containers do not need to withstand extensive transportation and storage, so often are less substantial than would otherwise be the case. Products purchased in larger quantities often need less packaging to deliver a given amount of product. By exhibiting a higher product-to-package ratio, they represent a form of source reduction.

- Retailers across the country offer dry food items from bulk delivery systems.
- A commercial bakery in Ontario reduced packaging waste 30 percent by switching from 50-kilogram bags of flour to 1-ton reusable cloth bags. The 1-ton bags are moved on skids by forklift and emptied directly into stainless steel hoppers.
- Restaurants and bars can serve beer and soft drinks on tap instead of in bottles and cans.

7. Combine Functions of More than One Product / Buy Multiple-Use Products

Combining the functions of two products into one may mean that less material is used. For example, a clock-radio or a toaster-oven may use less material than two separate appliances each designed to serve one of these functions.

- Combinations of household products—such as Bold with Fabric Softener, and Pert Shampoo and Conditioner—may use less packaging materials than if the two products are purchased separately.

8. Produce / Purchase Fewer Models or Styles

Producing and buying fewer product models can result in less turnover of useful products and therefore fewer resource and environmental impacts generated over the product life cycle. This option generally addresses trends in the proliferation of models and styles observed in a wide range of products. These changes often reflect consumer interest in fashion trends (for example, color or style) and may be of interest for source reduction efforts focusing on issues of lifestyle. This may be linked to Option 10—"Improve repairability"—because as the number of models proliferates, they become obsolete faster. Spare parts are less likely to be available, and it becomes more difficult to have the product repaired.

- The Volkswagen "Bug" and the Model T Ford offered few model options to the consumer but sold well and remained popular for many years.

9. Increase Product Life Span / Purchase Long-Lived Products

Making a product last longer, without a proportionate increase in the quantity or toxicity of materials used, will result in less material entering the waste stream. This can apply to increasing the durability of an appliance, producing a paint that lasts longer, developing a razor that does not have to be replaced as often, etc.

- Consumers should be encouraged to purchase higher quality products that cost more but last longer. This can also be cost-effective for the consumer. Examples are long-life light bulbs and higher quality tires.
- The Smith & Hawken Catalog produced a new awareness among consumers of quality tools designed to last far longer than typical gardening tools.
- In the past four years, IBM has reduced the number of components in its Selectric typewriter from the previous 2300 parts to 190. With the reduction in parts, reliability has gone up by a factor of 10.

10. Improve Repairability / Maintain Properly

Making a product easier and less costly to repair increases the probability that consumers will repair rather than replace it, thus prolonging its useful life. Improving repairability differs from increasing life span, because the former assumes that a part will eventually wear out. This option ensures that this component can easily be replaced or that the product can be fixed. (Increasing the life span would be making a part that lasted longer, without necessarily making it easier to replace when it did fail.)

- Manufacturers can franchise repair outlets for durable goods they create. This is still common with power tools and large appliances, but it has become less common with small appliances. They can also provide a toll-free number that consumers can call to find the location of authorized service centers for local repairs.
- Institutional consumers can purchase service contracts for equipment.
- Local governments and consumer groups can offer annotated directories of repair services available to consumers in their area. Consumers can consult the directory to find repair services more easily.

11. Produce / Purchase for Consumer Reuse

Producing products that can be used more than once, and purchasing and using such products, often results in source reduction. This option is frequently recommended because it is readily understood and offers an immediate alternative to other single-use products for which detailed life-cycle assessment information is not available. For example, reusable tote bags are often recommended as preferable to either plastic or paper bags. Examples are plentiful: purchasing reusable items such as pens, razors, and cameras instead of disposable ones; using mugs instead of any form of disposable cups; etc.

- Both natural and synthetic materials and innovative design features are now being used in the manufacture of consumer goods for reuse, such as shopping bags, lunch bags, microwavable containers, and thermos bottles.
- Gay Lea Foods of Weston, Ontario, started purchasing milk products in 1,000-gallon stainless steel containers, instead of the old 1,000-gallon

corrugated cardboard containers (which were reusable but had a plastic/foil insert that had to be discarded).

- Where feasible and safe to do so, consumers can reuse packaging such as plastic produce bags, paper bags, and glass jars for use in purchasing bulk foods from the store. They can also buy squeeze bottles and spray bottles once, and refill them from a larger recyclable container. For example, a Seattle food cooperative with 30,000 members encourages customers to reuse packaging at its six stores.
- Users of office equipment can purchase "recharged" cartridges for laser printers, copiers, and fax machines.
- Following a successful pilot project, Waste Management of North America and Goodwill Industries of Chicago jointly collect recyclable materials and reusable goods in Wheeling, Illinois. During the once monthly pickups, households can put out books, toys, small appliances, clothing, etc., for Goodwill, along with their newspapers and cans.

12. Produce / Purchase More Efficient Products

Generally thought of in terms of energy usage, efficiency can have broader applications involving materials use. The end result of efficiency is that less of a product is needed to provide the same service. Designing an aerosol nozzle to reduce overspray (so that more of the product reaches the target) would increase its efficiency. Reducing the solvent content of a paint, so that the solid-to-solvent ratio rises, could also be thought of as increasing efficiency.

- The hiding power of a paint (the number of coats that must be used to cover an underlying surface) depends on the pigments used. Manufacturers of household paints prefer to sell all the colors in a line at the same price. In order to do so, they sometimes cut back on the quantity of pigments used for colors where the pigment is significantly more expensive. This reduces hiding power, so that more coats of paint are used. Increasing the quantity of pigments in some paints would decrease the number of coats used, thus reducing the amount of solvent, resin and additives consumed in the end.
- Paper towel dispensers in restrooms that use rolled paper towels instead of folded ones allow users to take only as much towel as they need. When a nationally known theme/amusement park made this conversion, the Scott Paper Company estimated that it saved 210.5 tons of towels and 14.3 tons of towel packaging a year.

13. Use Product More Efficiently

In many cases, consumers can use a product more efficiently. For instance, there are various ways to use a piece of paper more efficiently: double-sided copying, using a smaller type (so that more words fit on a page), using the reduction feature on a copying machine (so that two pages can be copied onto one sheet), using the back for scratch paper, etc.

- Acu-Star Canada, Inc, a manufacturer of automotive parts, was purchasing boxboard to make patterns for cutting textiles. At the same time, they were disposing of boxboard they received as dividers in packaging. By using the packaging boxboard for their patterns, they reduced waste generation and saved $60,000 a year.

14. Produce / Purchase Complementary Products

Sometimes source reduction can be achieved by altering the characteristics or use of a companion product rather than the product being evaluated.

- Many refrigerators are manufactured with "rough-texture" exteriors to hide fingerprints, thereby reducing the need for cleaning (and thus reducing the use of household cleaners).
- Using latex instead of oil-based paint means that thinners do not have to be used for cleanup.
- Proper surface preparation may allow less paint to be used to coat a surface and may reduce the frequency of repainting.

15. Remanufacture Product / Purchase Remanufactured Product

Remanufacturing serves to extend the useful life of a product. It differs from repairing in that the product changes hands and is sold to a third party after being remanufactured. Examples include rebuilt carburetors and telephones.

- Goodwill, the Salvation Army, and similar organizations repair and resell donated items.

16. Borrow, Share, or Rent Product

There are many products that some consumers do not use frequently enough to require purchasing. By borrowing, sharing, or renting, they can save money, as well as conserve resources.

- An attractive feature of short-term equipment rental is that the rental business takes the burden of equipment maintenance from the renter.
- Neighborhood tool banks are becoming a popular means for household repair, remodeling. and maintenance tools to be shared by a number of people.

17. Other

The specific nature of a product category may be such that there are source reduction options, such as backyard composting, that do not fit under any of the choices in the checklist. This category is meant to account for such options.

Step 3
Select Options for Source Reduction

(Use Tool 3 — Evaluation Questions for Selecting Options)

Step 3 requires the analyst to evaluate source reduction options identified in Step 2 in order to select those options that are most effective in solving the problem(s) identified in Step 1. The information necessary to make this decision can be generated using Tool 3, "Evaluation Questions for Selecting Options."

Overview to Tool 3

The questions in Tool 3 are split into two parts, based on the type of information needed. (These are summarized in figure 2.9 and presented in full in the section that follows.) Part A contains questions designed to elicit *basic information* on the product that is likely to be useful for evaluating any of the options selected. The information used to answer these questions will *not* be used directly to select options but will serve as background information on the product. The overall goal of this section is for the analyst to understand the *product* well enough to be able to evaluate how effective any given option will be in solving the problem. In doing so, the information should also serve to pinpoint those product characteristics, models, etc., that are most pertinent for evaluation.

Part B contains questions that must be answered separately for each option identified in Step 2. The questions require *specific information* on each option. Part B is designed to evaluate the overall desirability of pursuing individual options, assuming that the policy strategies to implement it are 100 percent effective. These questions *will* be used directly in selecting options in Step 3. By determining which options appear to have the most potential, the analyst can then choose those options he or she wishes to evaluate further through Steps 4 and 5. (Manufacturers may decide to explore implementation of a promising source reduction option directly at this stage.)

Step 4 begins by identifying the obstacles to the options chosen in Step 3. It might seem logical to do this *before* choosing the options, with the rationale that one should not bother analyzing options that face major obstacles. However, doing so might lead one to abandon options prematurely. While conventional wisdom may indicate that there are insurmountable barriers, there may in fact be innovative solutions that would overcome them. Before dropping an otherwise promising option, the user should try to develop strategies to overcome the obstacles.

Finally, Tool 3 ends with a sample matrix (figure 2.10) intended to help facilitate decision making and comparison across options (see page 51). For each option under consideration, the analyst summarizes the response to each question in Part 3 as "unclear," "very negative," "negative," "potentially negative," "no effect," "potentially positive," "positive," or "very positive." Entering these values into the matrix makes it possible to "visually assess" a summary of the information on one page. Note that completion of such a matrix should not be a *substitute* for the detailed evaluation in preceding questions; it is a summary only.

As with the matrix in Tool 1, the purpose of this matrix is to simplify the information that has been gathered, and present it in a form where the trade-offs are more readily apparent. No weights have been developed to make "apples and oranges" comparisons, such as how to compare effects on product performance with effects on price. Users of the matrix will have to decide which options to choose based on their own subjective evaluation of the trade-offs between different questions.

How to Use Tool 3

Several general observations on the use of Tool 3 are offered. First, it is readily apparent that this tool requires substantial amounts of data. Manufacturers and key user groups are likely to be important sources of information and clarification of this data—particularly for the background questions in Part A. Second, although many questions do not specifically ask for trends information, changes over time can be significant for virtually any type of information

Figure 2.9
Summary of Tool 3 Evaluation Questions

Part A: "Basic Information for Any Option"

1. *Product*
   a. Models, types, sizes, or styles
   b. Quantity of the product
   c. Function of product or material of concern
   d. Toxicity
   e. Intended life of product and trends in average life span
   f. Economic life cycle and trends
   g. Major alternatives
   h. Market concentration
   i. Trends in product design, use, distribution, and pricing, and factors influencing those trends
2. *Source Reduction in the Industry / Firm*
   a. Source reduction experience by manufacturers
   b. Factors stimulating past or present source reduction
   c. Implications for future efforts
3. *Consumer Behavior*
   a. Consumer preferences
   b. Consumer misuse
   c. Source reduction experience by consumers
4. *Waste Management*
   a. Disposal patterns
   b. Recycling considerations
5. *Other*

Part B: "Information for Evaluating a Given Option"

1. *How This Option Applies to the Product and Its Potential Effectiveness for Reducing the Problem Identified*
   a. How much source reduction can be achieved? (based on general information obtained from Part A and an understanding of how this option applies to the product in question)
2. *Other Effects*
   What is the potential effect of this option on:
   a. Environmental trade-offs
   b. Product performance
   c. Product price and sales
   d. Manufacturers, retailers, distributors, and others
   e. Complementary products
   f. Alternative products
   g. Recycling and other waste management options
3. *Technical Barriers to This Option*
   a. Are there absolute physical or technical barriers that cannot be overcome?
4. *Other*

evaluated.

Third, answers to evaluation questions may eliminate from consideration certain options identified at the scanning step. It is even possible, though not likely, that the analyst will eliminate all options considered. (For instance, an analyst examining waste oil generation might fail to identify any feasible source reduction opportunities.) Such a situation could suggest serious informational gaps or perhaps difficulties in initial problem definition. In rare cases, it may suggest that other waste management options make more sense than source reduction.

Fourth, evaluation of one option or initial background questions may lead to the discovery of new options to consider. While the emphasis one places on certain questions depends on the types of options selected in Step 2 (for example, questions dealing with manufacturers versus consumer behavior), it may be worthwhile to explore questions that may not seem to be relevant initially.

Finally, note that questions within the tool do not need to be answered in the order given. The sequence will depend on how the information is collected and on the interests of the user.

Tool 3
Evaluation Questions for Selecting Options

PART A. BASIC INFORMATION FOR ANY OPTION

1. Product

a. Models, Types, Sizes or Styles, and Key Features that Differ

A target of analysis can often be subdivided into various classes, based on the functions it fulfills, performance characteristics, and materials used. This information can help the analyst focus on the types that are the most problematic. For instance, the product category "paint and coatings" can be divided into paint, varnish, stain, shellac, wood fillers and sealers, wood preservatives, specialty products, etc. It can also be divided in other ways, such as water-based and solvent-based formulations. Individual products can also be further subdivided. Paint can be characterized as interior or exterior, and flat or nonflat (satin, semigloss, or gloss).

The analyst will find it useful to identify different types of consumers and whether their consumption patterns vary by product type. For instance, painting contractors are more likely to use solvent-based paints than do-it-yourself homeowners. Two other important distinctions is whether consumers are institutions or individuals and whether they are the purchasers or end-users. (For instance, cafeterias purchase foodservice disposables, but the end users are those buying meals.)

b. Quantity of the Product

This information can provide both an indication of the amount of material that is disposed of and the economic importance of the industry. Various measures can be used, depending on the product and the interests of the analyst. Measures can include the number of items, weight, volume, sales volume (measured in dollars), etc. A given measure can also be expressed in various ways. For instance, weight can be measured in aggregate (billions of pounds) or per capita (pounds per person). If the "product" under consideration is a type of package, one measure of quantity is the amount of packaging per unit of product. It may also be important to link this question to question "a" above, by giving the quantities of different models, types. etc., or by user.

c. Function of the Product or Material of Concern

What are the desirable performance characteristics that the product or material of concern provides? For instance, if the product of concern is third-class mail, why do advertisers use this medium instead of newspaper, radio or television advertising? If the material of concern is the mercury in batteries, why is it used?*

Does the answer to this question vary for different products within a category and different actors? For instance, catalog companies may have different interests in using third-class mail than non-profits soliciting donations. And mercury serves different purposes in alkaline and mercury oxide batteries.

The quantity of the material of concern can also be disaggregated based on the different models or users

*It is also possible that a material of concern may serve no function. For instance, cadmium is a vital component of certain batteries, but it occurs in others as an impurity in the zinc that is used and serves no purpose.

discussed in question "a." In the case of batteries, how much mercury is used in alkaline batteries, mercury oxide batteries, and other batteries? How much mercury is used by household batteries, as opposed to those purchased for industrial, defense, or medical purposes?

To understand the function of the product or material of concern, it is important to understand how manufacturers view the performance characteristics of the product in general. There are generally numerous characteristics; those important to manufacturers may be different from the characteristics most important to consumers (although they determine the overall performance that consumers are interested in). For paint, the factors that manufacturers consider include durability (color fastness, wear resistance, stain resistance), hiding power (the number of coats it takes to cover a surface), adhesion to chalky surfaces, flow (ease of brushing), leveling (whether brush marks show in the surface), block resistance (not becoming sticky), surfactant leaching, sagging, appearance, drying time, odor, freeze-thaw stability, weatherability, corrosion resistance, resistance to scrubbing, ease of stain removal, and mildew resistance.

d. Toxicity

What toxic substances does the product contain? What are the effects, concentrations and length of exposure where those effects occur; what populations are at risk; etc.? What are the pathways of exposure during manufacturing, proper use, recycling, and disposal? What is the level of exposure from the life cycle of the product under consideration, including different methods of post-use management? Is there information from a quantitative risk assessment available? This information may be important in evaluating whether a change in product formulation will reduce toxicity. For example, one cannot know if substituting a different solvent in paint will reduce toxicity unless one understands why solvents present a risk, and what the impacts of both the original solvent and its substitute are.

e. Intended Useful Life of the Product and Trends in Average Life Span

If the product is disposable, how many uses is it intended to provide? If it is durable or non-durable, how long is its projected service life? Is there a difference between intended and actual life span based on the way a product is used (and, if so, why)? (This question may not be relevant to products that are consumed in use.) Aside from the simple issue of durability, life span is also of interest because it indicates how long it will be until products currently being purchased enter the waste stream.

f. Economic Life Cycle and Trends

Where is this product in the "economic life cycle" (that is, innovation stage, full maturity, gradually being replaced by alternatives)? Are markets expanding or declining? There is less concern about products that are in the process of being replaced by alternatives (unless the alternatives are worse). There is more concern about products where the market is rapidly expanding. There also are more opportunities to affect the design of products that are still in their innovative stage, because the industry has not yet standardized design.

g. Major Alternatives

What are the major alternatives to the product? If the material of concern is a toxic, are there less toxic or nontoxic substitutes available? Is it possible to totally eliminate the product or material, without substitutes? In the case of third-class mail, alternatives include private delivery, newspaper inserts, and advertisements in newspapers, magazines, billboards, radio or television.

h. Market Concentration

What is the market concentration of the product and its alternatives? Is information available by model, user, or other classes noted in question "a"? For food-service disposables, market concentration would be the share of the food-service market held by single-service implements and the share held by reusables. This information could then be analyzed for different users (that is, the market share of disposables in school cafeterias, nursing homes, hospitals, etc.) and for different products (trays, plates, flatware, etc.).

What is the market share of the firms manufacturing the product? Are there many firms or just a few? Are there industry leaders?

Are there consumers who buy a significant portion of a manufacturer's output, and thus have significant leverage over the manufacturer? For instance, a major fast-food restaurant chain can have a significant amount of influence over the type of products its suppliers produce.

i. Trends in Product Design, Use, Distribution, and Pricing, and Factors Influencing Those Trends

How are the factors in question "a" changing? Are there changes in the type of consumer or manufacturer? There may also be changes in the way a product is used or distributed. For instance, in the case of third-class mail there has been a trend from saturation mailing (sending material to as many people as possible) to targeted mailing (selecting recipients based on demographics).

What has been driving these trends? Factors include regulations, the cost of labor, product liability claims and insurance costs, socioeconomic factors (such as the shift from one-income to two-income families), fads, etc. It is also important to identify the actors involved with these issues.

2. Source Reduction in the Industry / Firm

a. Source Reduction Experience by Manufacturers

To what extent has source reduction been tried for this product—either in the United States or abroad? (Source reduction efforts abroad may offer suggestions or indicate trends in the United States.) What strategies were used? How long did it take to develop alternatives, and how much money was spent? What problems were encountered? How successful was the source reduction effort?

If there has been little source reduction in the past, then many avenues may be open. The challenge, then, is in overcoming the obstacles that have prevented source reduction from occurring. Conversely, past successes may mean that the easy steps have been taken, and future efforts will face greater barriers. Past failures can be as instructive as past successes.

b. Factors Stimulating Past or Present Source Reduction

What were/are the factors stimulating source reduction (cost savings, customer need, regulation, etc.)? Understanding the factors that have stimulated previous source reduction can be important in developing strategies to stimulate future source reduction. Regulations may concern disposal, or other aspects of the life cycle (for example, occupational exposure, safety, waste produced during manufacturing).

c. Implications for Future Efforts

What lessons can be learned from past and ongoing experience—that is, what do previous source reduction efforts suggest about opportunities for and/or impediments to future efforts? What strategies worked and did not work? Failure of past efforts may indicate the types of barriers that source reduction faces. If there have been notable success stories, they may show what types of motivating forces are effective.

3. Understanding the Role of Consumer Behavior

For the following questions, note that responses will vary depending on whether or not the consumer (or purchaser) is also the end user.

a. Consumer Preference

Why do consumers purchase this product? What are the key aspects that are important to them, including the product's attributes, packaging, and marketing? How closely is sales volume linked to price? (One method for gathering this information is through market research on consumer attitudes and perceptions.)

These preferences may also vary by type of consumer, or by the model, type, etc. of the product. For instance, painting contractors like solvent-based paint because of its appearance, workability, longevity, and the ease of surface preparation. They like latex paint because of its performance, ease of use/cleanup, and lack of odor. It may also be informative to examine how consumer preferences differ according to socioeconomic variables.

b. Consumer Misuse

To what extent is unnecessary waste caused by consumer misuse of the product? Consumer misuse of a product includes over-application; misapplication; improper disposal; ignoring directions, warnings, or service instructions; or mistreating the product.

c. Source Reduction Experience by Consumers

Have consumers reduced waste generation in the past by altering their purchases or habits? If so, what was the driving force? Have programs to influence consumers been tried in the past? With what results?

Do consumers know that they have choices, and how they can be acted on? For example, in the case of unsolicited third-class mail, do people know what they can do to stop receiving it? (This information can help identify obstacles to source reduction evaluated in Step 4.) Again, market research can provide information on how broadly and accurately consumers understand source reduction, what their attitudes are toward it, and how their actual behavior matches their attitudes.

4. Waste Management

a. Disposal Patterns

How is this product managed after its useful life? To what extent is the product "consumed" in use? For some products, only a fraction of the volume purchased is disposed of into the solid waste stream. The remainder may evaporate, be poured down the sewer, be discharged directly to the environment, etc. If this is the case, the product may be an important candidate for source reduction even if it does not have major impacts in the solid waste stream, since there may be serious environmental impacts through these other disposal routes.

b. Recycling Considerations

What are the most important options for recycling, how much is being recycled, and how does source reduction affect recycling? Are there waste management options that might be preferable to source reduction? Although recycling is lower in the waste management hierarchy than source reduction, it still has an important role to play for many products. Since source reduction activities can affect recycling, and vice-versa, it is important to understand the role of recycling and the trade-offs that may exist.*

While information on recycling may be helpful, recycling considerations should not be allowed to inhibit the consideration of source reduction options. The fact that a product is being recycled does not mean that source reduction is unnecessary. Source reduction is the preferred solution and should be pursued before recycling and other waste management options. (See chapter 1 for discussion of some problems associated with recycling.)

5. Other

Any other information that the analyst feels is relevant should be added here.

*Although source reduction may sometimes conflict with recycling, this is not necessarily the case. For instance, one can buy unbleached paper and copy double-sided, and then recycle.

PART B. INFORMATION FOR EVALUATING A GIVEN OPTION

1. How This Option Applies to the Product and Its Potential Effectiveness for Reducing the Problem Identified

Based on background information collected on this product and option, how effective might this option be? By how much will the quantity or toxicity of the waste be reduced? The potential depends on how widely the option is currently being used. (For instance, universal use of double-sided copying would not cut the use of copy paper in half, since many documents are already copied on both sides.) Will there be improvements elsewhere in the life cycle, in terms of materials, energy, and pollutants? If there is a life-cycle assessment available, it should contain this information (although it may be possible to answer this question without such an assessment). A life-cycle inventory will have information on quantities, while, depending on the option, additional analysis may be necessary to draw conclusions on environmental effects (see discussion in chapter 1).

2. Other Effects

a. Environmental Trade-offs

It is important to identify any negative impacts an option may have. There are many cases where a solution to one environmental problem will contribute to a different problem, creating an environmental trade-off. In the case of the paint and coatings industry, some manufacturers have replaced solvents such as mineral spirits, toluene, and xylene with 1,1,1 trichloroethane in order to reduce emissions of volatile organic compounds (VOCs). However, trichloroethane contributes to stratospheric ozone depletion, and may be classified as an air toxic.

Ideally, the analyst will have access to at least the first two stages of a life-cycle assessment. This will require an inventory of the types and amounts of inputs (such as raw materials, energy, water, etc.) and outputs (by-products released to air, water, and land) at each stage during processing, manufacturing, distribution and use, as well as an analysis of the resulting environmental effects. In practice, such complete information is rarely, if ever, available today.† Instead, the analyst will need to focus data collection on the

†This is especially important for toxics, with which there may be a switch to other chemicals whose environmental effects are less well known.

most significant negative impacts and to make qualitative assessments of any other impacts. Since precise information may not be available, estimates of the *magnitude* of the trade-offs will be important in making decisions.

b. Effect of This Option on Product Performance

If the option changes the product, or substitutes an alternative product, how does the performance of the replacement compare with what is currently being used? Does the replacement serve the primary function of the product category adequately? Does it serve secondary functions as well? Does the replacement have drawbacks in providing desired performance characteristics? How will these changes affect consumers?

c. Effect of This Option on Product Price and Sales

Is the alternative more or less expensive? The evaluation should also consider the time frame. Alternatives may have a higher purchase price, but cost less in the long run. On the other hand, they could cost more. For instance, a more durable item may cost more than several short-lived products once repair costs are factored in.

How closely is sales volume linked to price? This will affect both manufacturers and consumers. If the sales volume of a product does not vary significantly with price, manufacturers will be able to pass along to consumers any costs they may incur through reformulation or changes to the production process. If a price increase will lead to a large decrease in sales volume, manufacturers may be constrained in the alternatives they can adopt.

d. Effect of This Option on Manufacturers, Retailers, Distributors, and Others

Producing an alternative product may place a company at a competitive disadvantage with other firms in the industry, or it may put domestic manufacturers at a disadvantage compared to foreign manufacturers. Or, if consumers switch to alternatives, a different industry may reap the benefits.

There may also be advantages for producers. For instance, they may be able to gain a competitive edge by using materials more efficiently. Consumers may also be willing to pay a premium for products they perceive as being more environmentally sound.

If consumers and/or producers haven't adopted alternatives beneficial to them, then it is probably because of the existence of obstacles such as lack of information or the lifestyle changes required. But if these barriers can be overcome there may be a strong incentive for change.

e. Effect of This Option on Complementary Products

Would modifying this product produce an environmental improvement in complementary products (that is, those used in conjunction with this product)? For example, if the product category under consideration were paint-thinners and strippers, reformulating paint might allow different, less toxic, solvents to be used in the thinners and strippers.

Would modifying the product to protect the environment create incompatibilities with other products? For instance, would reformulating paint thinners, without first reformulating paints, mean that they would no longer work on certain types of paint? Or, could removing certain metals from batteries mean that they would no longer produce sufficient current to power certain appliances?

f. Effect of This Option on Alternative Products

If the option is implemented, what "spin-off" effects might there be for alternatives to the product? For example, if postage rates were increased for third-class mail, would advertisers turn to other media (e.g. newspaper, TV, radio) and with what environmental effects?

g. Effect of This Option on Recycling and Other Waste Management Options

Can source reduction be made compatible with recycling, recycled content, and safe waste management in such a way as to minimize impact on the environment? In some cases, certain source reduction options may conflict with recycling. For instance, switching from a package made of a single material to a multi-material may enable a manufacturer to reduce the package's weight and volume; however, the new package may not be as readily recyclable. Similarly, the methods used to make automobile tires more durable also make them more difficult to recycle.

Source reduction alternatives should not be promoted if they will interfere with other activities to such an extent that environmental impacts are increased. Instead, source reduction should always be structured as part of an integrated system to protect the environment.

Figure 2.10
Sample Matrix for Selecting Promising Source Reduction Options

Summary of analysis for product of concern / Options →	Eliminate product	Eliminate toxics	Substitute materials	"Lightweight" / reduce	Concentrate	Bulk / larger	Combine functions	Fewer models / styles	Life span	Repairability	Reuse	Efficienct products	Efficient use	Complementary products	Remanufacture	Borrow / share / rent	Other
1. *Effectiveness of source reduction option*																	
a. Overall effectiveness in solving primary problem																	
2. *Other effects from implementing the option*																	
a. Environmental trade-offs																	
b. On product performance																	
c. On product price and sales																	
d. On manufacturers, retailers, and distributors																	
e. On complementary products																	
f. On alternative products																	
g. On recycling and other waste management options																	
3. *Technical barriers to implementing the option*																	
4. *Other*																	

For options scanned in Step 2, give a value of **?** for unclear, **--** for very negative, **-** for negative, **-?** for potentially negative, **+?** for potentially positive, **+** for positive, **++** for very positive, and **/** for no effect.

3. Technical Barriers to This Option

Are there technical barriers that prevent an option from being adopted, or at least set an upper bound to the amount of source reduction an option can achieve? For example, there is a limit to how much a product can be concentrated; when all of the fluid is taken out of a liquid product it becomes a solid. Note that this question is meant to include only those barriers that are judged to be insurmountable in a scientific sense. Other types of technical or physical obstacles (such as the fact that a particular store does not have the space to store refillable bottles that customers return) that can be overcome (by tearing down a wall and enlarging the store) are addressed in Step 4.A, "Obstacles to Source Reduction Options."

4. Other

The above questions are not all-inclusive. If there are other factors that are important to an analyst, they should be added here.

Step 4
Scan the Range of Implementation Strategies

(Use Tool 4.A — "Obstacles to Source Reduction"; and Tool 4.B — "Checklist of Source Reduction Strategies")

The objective of this step is to identify potential policy strategies for implementing the source reduction option(s) selected in Step 3. Two tools are provided to help in that process.

Tool 4.A
Obstacles to Source Reduction Options

This tool is a discussion of various obstacles that can impede adoption of a source reduction option and corresponding questions the analyst should address. In summary, the suggested criteria are:

1. Technical obstacles
2. Information obstacles
3. Economic obstacles
4. Public policy obstacles
5. Consumer preference obstacles
6. Institutional obstacles

These obstacles are described in more detail below.

The need to first understand obstacles before selecting implementation strategies is based on the rationale that the option probably would have been implemented already, or implemented more successfully, were it not for the existence of such impediments as lack of information or consumer preferences. Strategies should be selected, therefore, according to their ability to overcome these types of obstacles. Tool 4.A is not intended to limit creativity in identifying potential strategies. Information gathered in this tool also will be used to compare the relative effectiveness of strategies in the next step.

As in the other tools, a response can fall under more than one heading. For example, consumer preference for a proliferation of models and styles may be a barrier to Option 8 in Tool 2 ("Produce fewer models or styles"). This can be seen as either a consumer preference obstacle or as an economic obstacle to manufacturers' adopting this option. What is important is that the obstacle is identified, not what heading it is placed under.

1. Technical Obstacles

Adoption of source reduction alternatives may be impeded by a variety of technical obstacles. More research and development may need to be done before alternatives are developed or commercially available. How much will this cost, and how long is it projected to take? If research needs are small, further study may be sufficient to solve this obstacle. If the best alternative will not be available for a number of years, it may make more sense to encourage the adoption of other alternatives in the interim.

The raw materials, production capacity necessary to switch to alternatives, etc., may not be available. How long will it take to overcome these obstacles, and at what cost?

There may be physical constraints that limit the adoption of alternatives. For example, a cafeteria may be unable to purchase reusable dinnerware because it lacks the space or fittings for a dishwasher; a particular store may lack the storage space for refillable bottles returned by customers. What will it cost to correct these obstacles, or what other accommodations will have to be made?

2. Information Obstacles

In some cases, manufacturers and consumers may not have adopted source reduction options because they lack the facts to make informed decisions. Their information may be incorrect, out of date, or simply unknown. The obstacles to obtaining information can be the proprietary nature of the information, the expense of obtaining the information, the lack of a means to

Figure 2.11
Selected Obstacles to Source Reduction

1. **Technical obstacles.** For example:

- More research and development is needed.
- Necessary raw materials and production capacity are unavailable.
- Physical constraints limit the adoption of alternatives.

2. **Information obstacles.** For example:

- Information necessary to make informed decision is proprietary.
- Necessary information is expensive to obtain.
- Current information is incomplete or faulty.

3. **Economic obstacles.** For example:

- Consumers would reject alternative as too expensive.
- Production of alternative product would place company at competitive disadvantage with other firms in the industry.
- Alternative is produced by a different industry.
- Profit margin or return on investment would be smaller for the alternative.
- Demand for alternatives is too small for manufacturer to produce them profitably.
- Firm lacks the funds for initial capital innvestment
- Producing alternative product could increase manufacturer's product liability costs.
- Waste generator is not responsible for full cost of waste disposal.

4. **Public policy obstacles.** For example:

- Government standards limit adoption of alternatives or force the substitution of less environmentally benign alternatives.
- Government policies (taxes, subsidies, import restrictions, etc.) provide contradictory incentives or otherwise present obstacles to adoption of alternatives.

5. **Consumer preference obstacles.** For example:

- Alternative requires a change in lifestyle or other behavioral changes.
- Alternative is less convenient to use.
- Consumers value the look or feel of a particular product.
- Consumers might refuse to buy product that is hard to find in the market place (e.g., available only through specialty shops or mail order).

6. **Institutional obstacles.** For example:

- No institutional structure exists to encourage source reduction.
- No mechanism exists to make individuals, organizations, or parts of organizations accountable for their actions.
- No infrastructure exists to support source reduction.

transmit the information. It is often the case that the necessary research has not been done, so that the information does not exist. For instance, there may be insufficient information on alternatives for toxics, or a life-cycle assessment needed to compare two products may not have been performed. Some examples of information that manufacturers and/or consumers may lack include the volume of waste generated and the cost of disposal; the problems that production, use. or disposal of this product creates; and the availability, performance, costs or savings, and size of the market for alternatives.

3. Economic Obstacles

There may be a variety of economic obstacles to a source reduction option. One common obstacle is that consumers reject the alternatives as too expensive. This can be due to the maintenance, repair or labor costs, as well as the purchase price.

Different economic obstacles can also impede the manufacture of alternative products. These can include: producing an alternative product would place a company at a competitive disadvantage with other firms in the industry; the alternative is produced by a different industry entirely; the profit-margin or return on investment is smaller for the alternative; market studies or experience show that consumers will not buy the alternative; demand for alternatives is too small for manufacturers to produce them profitably (it is below the minimum efficient scale); the firm lacks the funds for the initial capital investment required to begin producing the alternative; and producing an alternative product may increase a manufacturer's product liability costs (for instance, by making the product less resistant to product tampering).

Another economic obstacle can be that the waste generator is not responsible for the full cost of waste disposal, and thus does not have an economic incentive to include these costs in decision making. For example, senders of unsolicited mail do not pay for its disposal. School cafeterias may not pay for water, electricity or disposal costs out of their own budget. (This may also be an information obstacle, since no one may even know what these costs are, or an institutional obstacle, as noted below.)

4. Public Policy Obstacles

Government standards (for safety, environmental protection, etc.) may limit adoption of alternatives, or force the substitution of less environmentally benign alternatives. In some cases this takes the place of a trade-off. For instance, paint manufacturers may comply with regulations to control smog by replacing such solvents as mineral spirits, toluene, and xylene with 1,1,1 trichloroethane, which contributes to stratospheric ozone depletion and may be classified as an air toxic.

There are many cases where such standards are implemented for good reason. For instance, refrigerator doors using catches and flexible gaskets seal effectively, and are thus energy efficient. However, catches were replaced by magnetic seals to prevent children from locking themselves in and suffocating. Energy efficiency was reduced until more effective magnetic seals could be developed.

In addition to standards, other government policies (taxes, subsidies, import restrictions, etc.) may provide contradictory incentives or otherwise present obstacles to the adoption of alternatives. For instance, there is a limitation on imports of cotton diapers from China (designed to protect an American manufacturer of cloth diapers) that is making it difficult for diaper services to expand. Those who believe that cloth diapers are better for the environment than disposables would see this import restriction as creating an obstacle.

5. Consumer Preference Obstacles

There are many factors other than cost (which is covered in "Economic Obstacles") that affect consumers' purchasing decisions. In the absence of a strategy to encourage their purchase, consumers might not buy alternative products that do not meet these needs. For instance, the alternative may require a change in lifestyle, or there may be other behavioral changes necessary. It may take more time or effort to use, or be less convenient in some other way. There may be other intangible factors that consumers value, such as the look or feel of a particular product. Or they may refuse to buy products that are hard to find in the market place, such as those that must be purchased at specialty shops or through the mail. Market research may indicate what consumers think and whether an option will meet resistance or prove acceptable.

The increasing trend toward shorter product life spans ("planned obsolescence") reflects consumer preferences for new styles or for products that cost less than more durable alternatives and that they can replace rather than maintain or repair. Others might believe such trends are beneficial largely to manufacturers. In either case, consumer obstacles may be a factor in switching to product alternatives or modifying lifestyles.

6. Institutional Obstacles

Institutional factors can prevent manufacturers and consumers from implementing source reduction. Such obstacles include organizational structure, the lack of full-cost accounting, and the need for the necessary infrastructure.

The lack of an institutional structure to encourage source reduction may prevent it from happening. An organization, in both its capacity as a producer and a consumer, may not have a clear policy or guidelines on source reduction. There may be no one responsible for implementing such a policy, or sufficient resources may not have been devoted to it. A formal organizational structure to reduce the amount or toxicity of individual products, either in individual companies or in the industry as a whole, may serve to encourage source reduction.

Another institutional obstacle may be the absence of a mechanism to make individuals, organizations, or parts of organizations accountable for their actions by linking the costs or benefits of source reduction to purchasing and use decisions. This applies to both manufacturers and consumers. For instance, as has already been noted, the electricity, water, and waste disposal costs for school cafeterias are often paid for out of the school's general fund, instead of by the cafeteria itself. In fact, no one may even know what these costs are. Thus, the cafeteria may not have an incentive to conserve resources. Similarly, in communities where trash collection is paid for through property taxes or flat fees, as opposed to volume-based rates, consumers do not face the correct economic signals.

A final obstacle may be the lack of the necessary infrastructure to support source reduction. Some examples are: a product that could be redesigned to achieve source reduction, but the design changes would mean that most of the service personnel in repair shops around the country would need to learn how to repair the revised product; an organization that uses disposable products because no employee has the responsibility of collecting and cleaning reusable items; alternatively fueled cars that might not be driven outside of a major metropolitan area because there would be no facilities to refuel.

Tool 4.B
Checklist of Source Reduction Strategies

Tool 4.B. presents a checklist of strategies for implementing source reduction options selected in Step 3 (figure 2.12). Strategies fall into three categories: education, recognition and voluntary programs; economic incentives or disincentives; and administrative and regulatory actions. The three categories may often overlap, particularly in the case of administrative and regulatory actions (which can include education and economic incentives).

The following text includes a brief description of each type of strategy, including how it might affect source reduction. For example, a tax on raw materials would theoretically increase their cost and stimulate options that minimize the use of these materials in product manufacture.

Three factors should be considered when choosing a strategy: the ease of implementation, the distortions created, and the degree of effectiveness. Because intervention can create distortions, strategies should be chosen to minimize the amount of intervention necessary. It is only when a class of strategies will not be effective that a more interventionist strategy should be chosen. That is, education, recognition and voluntary programs should be considered before economic incentives, which in turn, should be considered before regulatory intervention. There are also gradations within classes of strategies. For example, within regulation, bans are more burdensome than mandatory disclosure of information.

No judgments are made by the Steering Committee regarding the relative merits or effectiveness of the specific strategies identified, with the general exception that, given a choice between strategies with an equal degree of effectiveness, the least interventionist strategy should be chosen.

As with the checklist in Tool 2, this scanning process can serve to eliminate strategies that clearly are not applicable to the options selected or it can highlight additional options for greater analysis. The analyst may wish to experiment with how comprehensive this initial scanning step should be. He or she may either wish to check off only those strategies that are clearly inappropriate, or select only a few of the most likely strategies given the product and source reduction option(s) selected.

Note that inclusion on the checklist is not intended

Figure 2.12
Tool 4.B — Checklist of Source Reduction Strategies

I. Education, Recognition and Voluntary Programs

__ A. Media / public outreach
__ B. In-store shopper awareness campaigns
__ C. School curricula
__ D. Services that promote source reduction activities
__ E. Instruction or training in source reduction techniques
__ F. Government dissemination of data on source reduction
__ G. Seminars or forums on source reduction
H. Awards
__ 1. Corporate awards
__ 2. Awards to individuals, including employees and product designers
__ 3. Awards to government agencies, educational or research institutions, or nonprofits
__ I. Labeling on environmentally preferable products
__ J. Stimulation of consumer demand for alternative products
__ K. Corporate guidelines for purchasing, operations, and sales
__ L. Corporate source reduction officers or offices
__ M. Corporate research and development into source reduction
__ N. Voluntary goals and standards for source reduction
__ O. Industry-wide product or packaging guidelines
__ P. Employee training
__ Q. Waste audits and product and packaging audits
__ R. Life-cycle assessments
__ S. Writing of letters or taking of other action to influence government and producers
__ T. Boycotting of producers
__ U. Other

II. Economic Incentives or Disincentives

A. Taxes or fees (with all revenues spent on source reduction, not going into general revenues or reducing the budget deficit) or tax credits, on the basis of activity, product, or material

Activities
__ 1. Volume-based disposal rates (variable can rates)
__ 2. Charges to landfill users based on full capital, operating, closure, and postclosure costs
__ 3. Landfill surcharges to provide revenues to finance source reduction projects
4. Tax credits to businesses conducting source reduction activities such as:
__ a. Product redesign, product line modification, purchase of source reducing equipment, research and development
__ b. Reserving specified amounts of retail shelf space for products designed for less and lower toxic waste
__ c. Specializing in the repair, restoration, or remanufacturing of products
__ d. Participating in efforts to standardize products to facilitate repairability and interchangability of parts
__ e. Conducting research and development on source reduction
__ 5. Other

Products
__ 6. Waste tax for every product sold, based on the quantity of waste it generates
__ 7. Tax on hard-to-dispose-of items (tires, waste oil, antifreeze, etc.)
__ 8. Tax on certain products that create environmental problems
__ 9. Tax on certain disposable products
__ 10. Deposits on certain products

__ 11. Tax rebates to consumers and/or manufacturers of reusable, repairable, remanufacturable, or more durable products
__ 12. Other

Materials

__ 13. Tax on virgin materials
__ 14. Tax on all raw materials
__ 15. Tax on certain toxic materials
__ 16. Marketable permits on certain uses of toxics
__ 17. End any existing federal tax subsidies for virgin material extraction
__ 18. Other

B. Direct Subsidies

__ 1. Government or industry grants to academia and nonprofits for source reduction research
__ 2. Loans or grants to stimulate initiation of source reduction pilot projects
__ 3. Incentives to individuals for source reduction
__ 4. Subsidies for source reduction infrastructure
__ 5. Low-interest loans or investment tax credits for businesses to purchase source reducing equipment or to make product line modifications necessary to produce alternative products
__ 6. Subsidies to businesses conducting source reduction activities
__ 7. Subsidies to consumers and/or manufacturers of reusable, repairable, remanufacturable, or more durable products
__ 8. Other

C. Administrative and Regulatory Actions

Activities

__ 1. Waste reduction officers
__ 2. Mandatory waste audits
__ 3. Change in regulated rate structures
__ 4. Elimination of restrictions or regulations that are obstacles to source reduction
__ 5. Government research and development on source reduction techniques
__ 6. Waste reduction planning requirements
__ 7. Ban on waste management options (such as landfilling certain items)
__ 8. Ban on certain activities
__ 9. Other

Products

__ 10. Procurement for source reduction by all levels of government
__ 11. Ban on government purchase of certain items
__ 12. Mandatory disclosure of virgin material use, energy use, air and water emissions, and solid waste produced
__ 13. Mandatory disclosure of product's estimated durability under "reasonable" use conditions
__ 14. Product durability (useful life) requirements
__ 15. Product reusability requirements
__ 16. Product constituent regulation
__ 17. Ban on products
__ 18. Ban on product attributes (disposable, multi-material)
__ 19. Other

Materials

__ 20. Mandatory disclosure of presence, amount, or concentration of toxics
__ 21. Product warnings for toxics (negative labeling)
__ 22. Restricted use of certain toxic substances
__ 23. Ban on materials
__ 24. Other

as an endorsement of any of the strategies. In fact, different strategies are strongly opposed by various Steering Committee members, depending on the specific circumstances. Judgments on strategies cannot be made independently of the product and option to which they apply; inappropriate uses of these strategies could actually increase life-cycle impacts from some products. (For example, banning certain products or materials could lead to the use of replacements with greater life-cycle impacts.) Tool 4B is designed only to describe the list of possible strategies. Evaluating the desirability of the strategies, as applied to the product and option under consideration, occurs in Tool 5.

I. EDUCATION, RECOGNITION, AND VOLUNTARY PROGRAMS

These types of strategies address gaps in knowledge or awareness that could otherwise encourage source reduction behavior by individuals or manufacturers.

A. Media / Public Outreach

Outreach can take the form of public service announcements, advertising, articles, shows, stories, etc., in a variety of media (newspapers, magazines, newsletters, TV, radio) to educate people about solid waste problems and the actions they can take to reduce waste generation. Examples include municipalities providing information on backyard composting techniques and manufacturers informing consumers about how to use a product more effectively.

B. In-Store Shopper Awareness Campaigns

Brochures, signs, posters, etc., at the point of sale can be designed to provide consumers with information they may need "on the spot" to make informed choices among products.

C. School Curricula

Educational material and programs can be designed to educate students on solid waste problems, and the actions they, government and businesses can take to solve them. In addition, college curricula for engineers and industrial designers can be revised to address source reduction techniques.

D. Services that Promote Source Reduction Activities

Such services could include running computer data banks to set up car pools and providing guides to goods and services, such as second-hand stores and repair shops.

E. Instruction or Training in Source Reduction Techniques

Explaining source reduction techniques, such as how to use a composting bin, conduct waste audits, etc., can serve to educate people about a particular technique and raise their general level of awareness. Instruction can take the form of booklets, classes, or one-on-one interaction.

F. Government Dissemination of Data on Source Reduction

Government agencies can provide information on source reduction in a variety of ways, including: toll-free numbers for people to ask questions; reference libraries; guides to repair, rental, and used goods services; catalogs and order forms for government publications on source reduction; outreach activities.

G. Seminars or Forums on Source Reduction

Seminars provide an opportunity to provide technical information to a specific type of audience, as well as for people to ask questions. Forums can bring together people with different backgrounds and experience, who would not ordinarily meet, to exchange information.

H. Awards

An awards program can recognize organizations that reduce the amount and/or toxicity of municipal solid waste, and/or conserve resources, through product design, production procedures, or other activities. Such a program can encourage source reduction by increasing awareness of reduction opportunities and providing incentive for action. Employee incentive programs can stimulate creativity and higher performance. Individual-recognition programs can also enhance awareness of source reduction needs and issues. (See chapter 3 recommendations for awards.)

I. Labeling on Environmentally Preferable Products

Labeling programs can be grouped into two types: "environmentally preferred" and "standard setting." Environmentally preferred programs award labels to products that are determined to be environmentally superior, relative to similar products. Standard setting programs can control the use of such ambiguous terms as "better for the environment." In either case, programs can encourage source reduction by helping consumers make well-informed choices, and providing an incentive to manufacturers. (See chapter 3 section on labeling.)

J. Stimulation of Consumer Demand for Alternative Products

In many cases, environmentally preferred alternatives are on the market today, and the challenge is in getting consumers to purchase/use them. This could be accomplished by advertising, giving away samples, etc.

K. Corporate Guidelines for Purchasing, Operations, and Sales

Corporate guidelines can achieve source reduction by setting standards for purchasing source-reduced products, by operating in ways that generate less waste, and by selling improved products.

L. Corporate Source Reduction Officers or Offices

Source reduction officers are individuals given the responsibility of implementing source reduction programs within an organization. Such officers can disseminate information and ideas to their colleagues and can serve as coordinators of activities often scattered throughout an organizational structure or individual program.

M. Corporate Research into Source Reduction

Through research and development, corporations can discover innovative new products or production techniques that can reduce waste generation while retaining profitability.

N. Voluntary Goals and Standards for Source Reduction

Setting goals for source reduction can provide a target to work toward and a yardstick for measuring progress. Goals can be numeric or qualitative. (See chapter 1 discussion on goals.)

O. Industry-Wide Product or Packaging Guidelines

Manufacturers in an industry might design source reduction guidelines that effectively address common needs and interests. They can also stimulate acceptance by firms that might not otherwise have participated.

P. Employee Training

On-the-job training programs can teach employees how to recognize and implement source reduction alternatives in their organization, in its capacities both as a manufacturer and as a consumer.

Q. Perform Waste Audits and Product and Packaging Audits

Audits are a tool that can be used by businesses, government facilities, and other institutions to inventory and trace input materials and to identify waste generation patterns. With the information provided by audits, organizations can effectively target specific strategies for waste reduction. Audits enable organizations to identify changes that can be made in three general areas: procurement, operations, and end products.

R. Life-Cycle Assessments

Information from the first two stages of a life-cycle assessment (inventory and analysis of effects) may serve to point out environmentally preferred alternatives. They can also identify priorities for improvements in a particular product, by pointing out where the environmental weak spots are in its life cycle. This can be followed by the third stage of a life-cycle assessment (an analysis of the changes needed). (See chapter 1 and appendix A for more discussion of issues surrounding life-cycle assessments.)

S. Writing of Letters or Taking of Other Action to Influence Government and Producers

Individuals or groups can write letters, make phone calls, circulate petitions, and take other actions to persuade government or manufacturers to make changes that will benefit the environment.

T. Boycotting of Producers

Boycotts are intended to bring about changes in the characteristics of environmentally damaging products. This could entail refusing to purchase a particular product or all of a manufacturer's products.

U. Other

II. ECONOMIC INCENTIVES OR DISINCENTIVES

A. Taxes / Fees / Credits

Many environmental costs associated with producing or consuming a product are not included in market prices and are therefore called "externalities." Demand is higher due to these low prices, and hence the polluting activity—amounts and toxicity of solid waste—is higher. The general rationale behind the tax strategies listed below is that the price better reflects the environmental costs, and thus individuals or entities are forced to account for them. They also allow a freedom of choice unavailable under regulatory strategies. Note that in some cases it would be possible to fashion an incentive to correspond to the taxes (disincentives) listed below.

Activities

1. Volume-Based Disposal Rates (Variable Can Rates)

Volume-based disposal rates charge households for trash pickup on the basis of the number of cans or bags set out, instead of a flat fee or through property taxes. This gives households a direct economic incentive to reduce household and yard waste.

2. Charge Landfill Uers Full Capital, Operating, Closure and Postclosure Costs

In the absence of such a requirement, many municipal landfills may charge users only for operating costs and pay for other costs out of general tax revenues. This artificially lowers the price of landfilling, which in turn can retard source reduction.

3. Landfill Surcharges to Provide Revenues to Finance Source Reduction Projects

Such surcharges assume that there are externalities elsewhere in the system. Because it may not be desirable to use an incentive to adjust pricing before the waste is generated, this program uses revenues generated from disposal to fund programs to reduce that waste.

4. Tax Credits for Businesses Conducting Source Reduction Activities

Tax credits can serve to stimulate activities such as product redesign, product line modification, purchase of new equipment, standardization of products and parts, research and development, research and development in source reduction, etc.

5. Other.

Products

6. Waste Tax for Every Product Sold, Based on the Quantity of Waste It Generates

Such a tax is similar in rationale to volume-based disposal rates, but the charge is levied at the point of purchase, instead of disposal.

7. Tax on Hard-to-Dispose-of Items (Tires, Waste Oil, Antifreeze, Etc.)

Such items cause special environmental problems upon disposal and must be handled separately. A tax forces users of the products to pay the costs of these special handling programs, instead of society at large. They will theoretically reduce their consumption when the price increases to reflect true social cost.

8. Tax on Certain Products that Create Eenvironmental Problems

A tax on certain products that create environmental problems would promote the use of alternatives. This is similar to the tax that has been levied on chlorofluorocarbons (CFCs) to stimulate the production of alternatives that are less destructive to the ozone layer.

9. Tax on Certain Disposable Products

A tax on such products would raise their prices and provide an incentive to use reusable alternatives.

10. Deposits on Certain Products

Deposits can bring back tires for retreading, auto batteries for refilling, bottles for reuse, etc.

11. Tax Rebates to Consumers and/or Manufacturers of Reusable, Repairable, Remanufacturable, or More Durable Products

Such products often cost more to purchase, although they may be cheaper in the long run. Still, consumers may not be able to afford the "up-front" costs. Tax rebates to consumers would lower the cost the buyer pays, while rebates to manufacturers would allow them to reduce the sales price. In either case, the result would be to stimulate purchase of the products.

12. Other

For example, providing greater tax incentives for donating used goods to charitable organizations such as the Salvation Army and Goodwill could encourage donations and increase reuse of products.

Materials

13. Tax on Virgin Materials

A tax on virgin materials or products produced from virgin materials would increase their price. This would provide an incentive for manufacturers to use recycled materials, and to use less material per unit of product where virgin materials are needed, and could reduce consumer demand for products and materials. It could be applied when raw materials are sold to fabricators or at the point of sale of the final product, based on recycled content.

14. Tax on All Raw Materials

Such a tax would raise their price and provide an incentive to manufacture and consume products that use less material, thus providing a stimulus for source reduction.

15. Tax on Certain Toxic Materials

Such a tax would encourage the elimination or reduction of selected toxic materials and stimulate efforts to develop substitutes.

16. Marketable Permits on Certain Uses of Toxics

Such permits would set limits on the total quantity of a toxic material that could be used, and then allocate the rights based on market forces. This could be accomplished either by auctioning the permits or by allowing firms to buy and sell them. This would restrict the overall quantity of material used but allocate it to the most valuable uses.

17. End Any Existing Federal Tax Subsidies for Virgin Material Extraction

Extraction industries — such as mining, oil, and timber — historically have received various forms of tax subsidies, in the form of depletion allowances, favorable capital gains treatment, below cost leases, etc. Such subsidies lower the price of these materials and theoretically increase their use.

18. Other.

B. Direct Subsidies

Whereas taxes attempt to better incorporate the full cost of externalities, subsidies are payments to encourage activities that might not otherwise occur because of inadequate market prices or other lack of private incentive to invest in projects benefiting the public in general.

1. Government or Industry Grants to Academia and Nonprofits for Source Reduction Research

In many cases, attempts to choose source reduction options are hindered by a lack of information on issues ranging from technical performance of alternatives to consumer behavior. Funding the basic research on such questions provides the background necessary to develop options and strategies.

2. Loans or Grants to Stimulate Initiation of Source Reduction Pilot Projects

Innovative source reduction methods must be tested in real-life situations before they will be widely adopted. Pilot projects offer an opportunity to work out the bugs in a new approach.

3. Provide Incentives for Source Reduction to Individuals

This involves providing direct benefits to individuals to take actions that reduce waste generation, such as taking a few cents off their grocery bill if they provide their own grocery bags.

4. Provide Source Reduction Infrastructure

Providing source reduction infrastructure can be giving out or otherwise funding the actual equipment needed such as composting bins and cloth shopping bags.

5. Low-Interest Loans or Investment Tax Credits for Businesses to Purchase Source-Reducing Equipment or to Make Product Line Modifications Necessary to Produce Alternative Products

Although such source reduction activities have social benefits, they may not be profitable enough for firms to initiate them without financial help.

6. Subsidies for Businesses Conducting Source Reduction Activities

Subsidizing activities such as product redesign, product line modification, purchase of new equipment, standardization of products and parts, and research and development may act to promote source reduction.

7. Subsidize Consumers and/or Manufacturers of Reusable, Repairable, Remanufacturable, or More Durable Products

Such products often cost more to purchase, although they may be cheaper in the long run. Still, consumers may not be able to afford the "up-front" costs. Government payments to consumers would lower the cost the buyer pays, while grants to manufacturers would allow them to reduce the sales price. In either case, the result would be to stimulate purchase of the products.

8. Other.

III. ADMINISTRATIVE AND REGULATORY ACTIONS

Activities

1. Waste Reduction Officers

See discussion under "Corporate source reduction officers or offices" in section I.L above.

2. Mandatory Waste Audits

Manufacturers could be required to conduct waste audits. One option would be to require estimates of waste production to be made at the product design stage.

3. Change Regulated Rate Structure

In cases where governments regulate or set rate structures, such as mail, water, sewer, electricity, and gas, they can address externalities by changing prices instead of using other economic incentives. Such changes could be tied to certain conditions, such as requiring the use of unbleached envelopes to receive favorable third class-mail rates.

4. Eliminate Regulations or Other Activities that Are Obstacles to Source Reduction

As described in Tool 4.A, part D ("Public Policy Obstacles"), government policies may have unintentional side effects that can create barriers to source reduction. There may be cases where it would be appropriate to change the policy to remove this unintended barrier.

5. Government Research and Development on Source Reduction Techniques

Agencies may find it appropriate to carry out research and development themselves, instead of paying others to do it.

6. Waste Reduction Planning Requirements

Large-quantity waste generators can be required to submit annual reports indicating the quantity of waste generated, the reduction that is planned, and how they will achieve such reductions. One option is to require users of large quantities of toxic substances to submit reports on their planned toxics reductions.

7. Ban on Waste Management Options (such as Landfilling Certain Items)

If waste disposal options are narrowed, the cost of disposal rises, providing an incentive for consumers to reduce their purchases of a product.

8. Ban on Certain Activities

Certain activities that increase waste generation, such as multiple mailings to the same address by third-class mailers, could be prohibited.

9. Other.

Products

10. Procurement for Source Reduction by All Levels of Government

Federal, state, and local governments, and their contractors, purchase large quantities of goods. Procurement guidelines can be developed to encourage the purchase of appropriate equipment and supplies. Reducing the volume or toxicity of the waste governments and contractors generate would have a significant impact on the waste stream. Using their buying power, they might be able to influence manufacturers to produce new products not currently on the market. This would benefit other consumers as well.

11. Ban on Government Purchase of Certain Items

Such a program would prohibit the purchase of certain environmentally damaging items, such as inks containing heavy metals. It would operate through the procurement process, and its effect would essentially be the same as Strategy 10, above.

12. Mandatory Disclosure of Virgin Material Use, Energy Use, Air and Water Emissions, and Solid Waste Produced

It could be required that certain products be labeled with information on key environmental information, similar to energy efficiency labels currently used on appliances. This information, which is necessary to make life-cycle comparisons, would be gathered through a life-cycle assessment inventory. (See the definition of source reduction and discussion of life-cycle assessments in chapter 1 and appendix A for further details.)

13. Mandatory Disclosure of Product's Estimated Durability under "Reasonable" Use Conditions

A requirement that certain products be labeled with their estimated life-span, based on testing information, would allow consumers to make informed choices on durability.

14. Product Durability (Useful Life) Requirements

A product durability requirement would guarantee that a product would last for a specified period of time, with the manufacturer required to repair or replace any unit that failed before then. Thus, it would be similar to the warranties currently available for many products. It could reduce waste generation if it encouraged manufacturers to produce products that were more durable or more easily repaired.

15. Product Reusability Requirements

It could be required that packaging and other nondurable goods be manufactured to be reusable. This might take place through a change in material (switching from metal food packaging to glass) or by a change in design to make a product more durable and thus reusable (pallets, drums, boxes, etc.).

16. Product Constituent Regulation

A maximum allowable quantity or concentration of certain toxics in products could be established.

17. Ban on Products

The sale of a product could be prohibited.

18. Ban on Product Attributes (Disposable, Multi-Material)

The sale of products with a certain attribute (disposable, multi-material, etc.) also could be prohibited.

19. Other.

Materials

20. Mandatory Disclosure of Presence, Amount, or Concentration of Toxics

Requiring that products be labeled with information on toxics content, similar to nutrition labeling on food, would provide consumers with information necessary to make choices.

21. Product Warnings for Toxics (Negative Labeling)

Products containing certain toxic components could be required to carry a warning of possible health or environmental effects, similar to those on cigarette cartons. Such a warning would provide information to consumers to make choices.

22. Restricted Use of Certain Toxic Substances

Use of specified toxic substances could be restricted to certain conditions, similar to pesticide restrictions that now exist.

23. Ban on Materials

Prohibiting the use of a material of concern could accomplish some source reduction.

24. Other

Step 5
Select Policy Implementation Strategies

(Use Tool 5 — Evaluation Questions for Selecting Strategies)

The analyst's task at this step is to choose policy strategies to implement the desired source reduction options in a desirable and effective manner. Tool 5, "Evaluation Questions for Selecting Strategies," helps to generate information to assist in that decision. (If multiple options were chosen in Step 3, the analyst will need to work through this section separately for each one.)

The tool presents a series of questions addressing several different aspects of the strategy (figure 2.13). To answer these questions, the analyst should think through how this strategy would work for the option under consideration and determine whether more than one strategy may be needed to implement an option. For instance, backyard composting can be promoted by a direct subsidy (giving away composting bins) in combination with education (training people how to use them). When two or more strategies must be combined to be reasonably effective, they should be evaluated as an integral package.

Note that the general guidelines and caveats for Step 3 apply here as well. First, note that this is an iterative process. Information gleaned through the evaluation process may lead to the discovery of new strategies to consider. Second, the amount of effort spent on certain questions will vary depending on which strategies were selected in the previous step.

Finally, Tool 5 ends with a sample matrix intended to help the analyst summarize information collected and to compare various strategies assessed (figure 2.14). As in Step 3, the analyst would assign a value of "unclear," "very negative," "negative," "potentially negative," "no effect," "potentially positive," "positive," or "very positive" to each question. These values could then be entered into the matrix.

A separate matrix must be prepared for each option under consideration. Also, if more than one strategy is needed to implement an action, they should be entered on the matrix as parts of a "package," and the analyst should make a choice between different packages in Step 5.

As with the matrices in Tools 1 and 3, the purpose of this matrix is to simplify the information that has been gathered and present it in a form where the trade-offs are readily apparent. No weights have been developed to make "apples and oranges" comparisons, such as how to balance social equity against economic efficiency. Users of the matrix will have to decide which strategies to choose based on their subjective evaluation of the trade-offs between different questions.

Tool 5
Evaluation Questions for Selecting Strategies

1. How Does This Strategy Apply to the Option, and Would It Effectively Overcome the Obstacles Identified?

Based on background information collected on the obstacles and knowledge of how this strategy would work, how effective might this strategy be in overcoming the obstacles and implementing the option? Is it likely to overcome the most important obstacles impeding this option? Does it require companion strategies to address key obstacles? (If so, reevaluate as a package.)

How effective will the option be at solving the primary problem, given the effectiveness of this strategy? For example, if an option had the potential to reduce the waste stream associated with a product by 5 percent if the strategy implementing it were 100 percent effective, but the strategy chosen realized only 50 percent of the potential, then the option and strategy would reduce this waste stream by 2½ percent.

Figure 2.13
Summary of Tool 5 Evaluation Questions

1. **Application of this strategy to the option and its potential effectiveness for overcoming the obstacles identified**

2. **Feasibility of implementation**
   a. Actors to implement
   b. Resources
   c. Access to information

3. **Burden associated with the strategy**
   a. Least amount of intervention
   b. Social and economic equity
   c. Economic efficiency

4. **Other**

2. How Feasible Is Implementation of This Strategy?

a. Who Are the Actors to Implement It?

How probable is it that a specific institution could readily adopt the role required for an option to be implemented in a sound manner? This will depend on factors such as whether the institution has the necessary trained personnel, how closely the task fits within its missions, whether it has experience administering similar strategies, legal authority, etc.

b. What Resources Are Necessary for Implementation?

How large is the commitment of resources necessary to implement this strategy, in terms of staff and money? (For matrix response, minimal or no resources would be very positive.)

c. How Accessible Is the Information Required for Implementation?

Is the information or data required to implement this strategy available at a reasonable cost and within a reasonable amount of time?

3. How Burdensome Is This Strategy?

a. Does This Strategy Involve the Least Amount of Intervention Necessary to Accomplish the Task?

How disruptive would the strategy be to the current market system? In the absence of externalities, the free market system works most efficiently with a minimum of government intervention. But externalities do exist, creating the need for government action. Because intervention can create distortions, strategies should be chosen to minimize the amount of intervention necessary to correct externalities. It is only when a class of strategies will not be effective that a more interventionist strategy should be chosen. For instance, economic incentives should be chosen only if they are significantly more effective than education, recognition, or voluntary programs.

b. Is This Strategy Socially and Economically Equitable?

Source reduction strategies are likely to have different impacts on different segments of society. Are there disproportionate burdens on any group?

c. What Is the Economic Efficiency of This Strategy?

To be economically efficient, the total social benefits of an action must exceed the total social costs. In addition, the strategy should be cost-effective (that is, achieve the desired results at the lowest possible cost). Performing such a calculation is not a trivial exercise. It will rarely, if ever, be possible to collect complete information on the benefits and costs of an action. However, this does not argue against the concept of comparing total benefits with total costs. (Complete information on toxicity is often unavailable, but this does not prevent toxicity from being considered in decision making.) Instead, it means that the analyst

Figure 2.14
Sample Matrix for Selecting Implementation Strategies

Strategies →	Strategy Package 1			Strategy Package 2			Strategy Package 3		
Summary of analysis for product of concern	Strategy A	Strategy B	Strategy C	Strategy A	Strategy B	Strategy C	Strategy A	Strategy B	Strategy C
1. Effectiveness of strategy									
a. Overall effectiveness in overcoming obstacles									
2. Feasibility of implementation									
a. Actors to implement									
b. Resources									
c. Access to information									
3. Burden associated with the strategy									
a. Least amount of intervention									
b. Social and economic equity									
c. Economic efficiency									
4. Other									

For strategies scanned in Step 4, give a value of **?** for unclear, **--** for very negative, **-** for negative, **-?** for potentially negative, **+?** for potentially positive, **+** for positive, and / for no effect.

*"Strategy Package" refers to a set of strategies that, if taken together, could be used to implement a source reduction option. Of course, some strategies may be effective in and of themselves without complementary actions.

must do the best job possible with the available information. It is important to understand the limits of the methodology and the imprecision of the estimates when interpreting the results.

4. Other

The questions identified above are meant to be useful but are not necessarily all-inclusive. Users are encouraged to modify them to suit their own particular needs, or to add further questions.

Chapter 3
Selected Strategies to Encourage Source Reduction

The previous chapter presented an Evaluation Framework that decision makers in various sectors can use to help select effective source reduction options and implementation strategies. This chapter takes a closer look at several types of strategies: actions that can be undertaken without the need for substantial new research, a national awards program, and product labeling programs. The chapter concludes with an assessment of research needs and other recommendations that should be pursued for signficiant progress to be made in source reduction.

THINGS WE CAN DO NOW

There are numerous types of strategies that require no further study or complex assessment before action can be taken. Substantial amounts of source reduction can be accomplished *now*. Several examples are provided below to illustrate just a few of the actions that can be taken by individuals, businesses, and governmental entities.* In addition, Tools 2 and 4 in the preceding chapter contain examples of things manufacturers and consumer *are* doing now.

Source Reduction in Your Own Back Yard

Yard wastes are a large component of municipal solid waste (MSW) in the United States. These grass, leaf, and other yard clippings comprise about 18 percent of the total amount of wastes generated and about 10 percent of the volume of wastes that require disposal.[1] While landfilling yard debris uses up disposal capacity and contributes to methane gas, leachate, and settling problems, incineration of yard wastes contributes nitrogen oxide emissions and is inefficient because of the waste's high moisture content. Backyard composting offers an attractive solution to these problems and produces a beneficial mulch for landscaping in the bargain.† Successful programs to initiate backyard composting make composting as easy and inexpensive as possible, since residents often may not have the knowledge or economic incentives to undertake composting on their own.[2]

- Seattle has been educating residents on the best methods for backyard composting program since 1986. In December 1989, Seattle began distributing free compost bins through its Backyard Composting Program, with a goal of distributing 70,000 bins by 1998. In its first year, the program provided 10,800 free bins to residents, along with an optional one-hour personalized instruction on composting. First-year program costs totaled $675,000 for the bins (at about $19 each), salaries for composting trainers, advertising, and other administrative costs. Benefits are impressive. When all the bins are distributed, about 10 percent of Seattles's annual residential yard waste will be diverted from the municipal waste stream at a savings of nearly $3 million dollars over the next 20 years.[3]

Charging by the Bag: Volume-Based Disposal Rates

Most consumers do not face economic incentives to reduce household and yard wastes. More often than not, consumers pay for trash collection indirectly through property taxes or by flat fees regardless of the amount of material left for pick up. Strategies that charge households and commercial enterprises on the

*Readers are encouraged to contact local sources to identify source reduction programs in their area. Also, see the bibliography for more information.

†In contrast, composting operations in which yard wastes are collected by municipal entities are *not* considered to be source reduction because the material has already entered the waste stream for management.

basis of quantity, known as volume-based rates or variable-can rates, *can* provide a straightforward incentive to change behavior. An increasing number of communities are experimenting with such schemes and finding them successful. Whatever system is used, the challenge is to set the charge high enough to cover intended costs and provide the "correct" market signal but low enough to discourage illegal attempts to avoid the system.* Major metropolitan areas, such as Seattle, are not alone in adopting volume-based rates; smaller towns and cities are as well.

- Residents of Perkasie, Pennsylvania buy 20- and 40-pound size trash bags from the borough at a cost of $0.80 and $1.50, respectively. There is no charge for the collection of recyclables. In the first six months of the program, the amount of trash collected (both wastes and recyclables) fell by 28.7 percent.[4]
- High Bridge, New Jersey, sells stickers that must be placed on trash bags or on the top item in a trash can. Fifty-two stickers (one for each week of the year) cost about $2.70 each with additional stickers costing $1.25 each. Recyclables are picked up separately. Total tonnage of garbage has been reduced 24 percent as a result of the program.[5]

Changing Office Practices

Personnel in most businesses, government, and other offices often are not aware of the waste associated with routine office practices. Many local and state governments (as well as commercial firms) are beginning to offer technical assistance, in the form of waste audits, to evaluate waste generation patterns and to target specific strategies for improvement.[6] While waste audit programs traditionally have focused on waste management and recycling improvements for small and medium-sized industries, they are reaching out increasingly to commercial and other offices.

Sometimes the recommended strategies are simple: copy documents on both sides, use backs of single-sided copies for scrap paper, and use mugs rather than disposable cups. Even in these cases, however, systematic appraisal of office practices and proper design of the program can make the difference between success or failure.

*Attempts to avoid the collection system can be minimized by helping customers reduce their charges (that is, by educating consumers on source reduction measures they can use at home and the office and by and by providing recycling and composting programs).

Changing Procurement Practices

Waste audits often lead to recommendations for changing procurement practices that can contribute significantly to waste reduction. For example, audits can help an establishment determine whether the packaging it is using contains toxic elements, whether the paper it is using contains recycled fibers, and whether the supplies it is using are excessively packaged. Consequently, it can request suppliers to provide more appropriate materials.

Additional procurement options include purchase of highly durable equipment, use of service contracts, purchase of reusable products over disposables, investment in computerized mail software, and careful evaluation of necessary quantities for supplies and documents. While initial investment in these purchases may be substantial, savings can materialize in the longer run due to the reduction in waste.

Educating for Source Reduction

Many of the best opportunities for source reduction are through individual purchasing and disposal decisions. A wealth of guides are emerging to help guide consumer purchases and behavior.[7] State and local governments are adopting a variety of materials for use in schools, including curricula on solid waste, source reduction, and consumer choices. A compendium of state programs on solid waste education, including source reduction, has been compiled,[8] while the U.S. Environmental Protection Agency (EPA) has published a bibliography on solid waste curricula.[9] Two examples of state educational programs include:

- Ohio requires that all schools include environmental education in the curriculum and has developed a comprehensive solid waste, recycling and litter prevention curriculum guide.[10]
- Rhode Island has developed a two-volume solid waste curriculum for grades 4-8 that includes a chapter on source reduction.[11]

Reducing Waste from Unsolicited Mail

Unsolicited mail is one of the fastest growing segments of municipal garbage today. Third-class mail alone weighed 3.3 million tons in 1986, or about 1.5 percent of municipal waste,[12] and this excludes unsolicited mail sent in first and second classes. Not only does unsolicited mail generate sizeable amounts of waste, but it does so for very little consumptive value for most recipients. The U.S. Postal Service estimates that 93 percent of all third-class mail is discarded the day it is received. Several courses of action are available:

- Consumers can request all direct mailers with which they do business to not have their names sold or traded to other mailers.
- Consumers also can terminate mail already being received by requesting those companies to remove their names from their mailing lists.
- Individuals and institutions can write to the Mail Preference Service of the Direct Marketing Association (DMA), asking that their names be added to a "delete file" which is made available to businesses on a quarterly basis.* Names are maintained on file for five years but must be re-registered in the event of a change of address. Consumers may register their names, including spelling variations with DMA, by writing to:

Mail Preference Service
Direct Marketing Association
P.O. Box 3861
11 West 42nd Street
New York, NY 10163-3861

A NATIONAL AWARDS PROGRAM FOR SOURCE REDUCTION

The Steering Committee recommends that an annual national awards program be established to promote reduction in the amount and toxicity of municipal solid waste produced. An awards program could raise awareness of source reduction opportunities and provide incentives for manufacturers to design and implement innovative production procedures and products.

The program should have the following general characteristics: 1) it should recognize a variety of organizations that have undertaken outstanding source reduction initiatives, 2) criteria used to determine award winners should include environmental merit, innovativeness, transferability, economic value, and commitment to environmental protection, and 3) an independent panel of experts should determine award winners, with technical assistance provided as needed by an independent testing organization(s).

Recommendations

1. Encourage Diverse Participation

The Steering Committee recommends that an awards program recognize a variety of groups, including manufacturers, commercial enterprises, public-interest groups, state and local government agencies, and educational institutions that make an outstanding contribution to source reduction.

2. Recognize Specific Categories of Achievement

Separate awards categories should be established for different types of organizations. For manufacturers and perhaps other groups, award categories should be further narrowed to recognize explicitly various types of innovative source reduction activities. For example, one award category could be "development by a manufacturer of an innovative packaging design." Below is a list of eight categories identified by the Steering Committee that might be utilized in a comprehensive source reduction awards program:

- Development and/or implementation by a manufacturer of an innovative *product design* that reduces the amount and/or toxicity of waste produced;
- Development and/or implementation by a manufacturer of an innovative *packaging design* that reduces the amount and/or toxicity of waste;
- Development and/or implementation by a manufacturer of an innovative *production procedure* that reduces the amount and/or toxicity of waste;
- Outstanding contribution to source reduction by a *service firm or commercial enterprise*;
- Outstanding contribution to source reduction by an *educational or research institution*;
- Outstanding contribution to source reduction by a *state government agency*;
- Outstanding contribution to source reduction by a *local government agency*; and
- Outstanding contribution to source reduction by a *public interest group.*

Since the goal of the proposed awards program is to promote source reduction of MSW, awards should be given only for activities that directly further this end. Separate award programs might address other types of waste management activities such as recycling, but source reduction of MSW should be the focus of this program.

Each source reduction award should be attached to a specific activity performed or product produced within a specific period of time. The presentation of an award to an organization should not represent a blanket approval of all of an organization's activities and products.

*It may take several months before there is a noticeable decrease in the mail received through these marketing lists.

3. Use Several Different Criteria

Environmental Value

The ultimate goal of a source reduction awards program is to promote a cleaner environment by stimulating reduction in the amount and toxicity of municipal solid waste produced. Therefore, the most important criterion in determining award winners must be an assessment of the environmental benefits associated with various activities or products.

Estimates should be made as to how much various activities or products reduce the amount and toxicity of MSW. In addition, some consideration must be given to other environmental impacts of activities in order to ensure that the overall environmental impact of an award-winning activity or product is positive. For example, a change in product design could reduce the amount and toxicity of waste produced but be much less energy efficient. Studies that take into consideration life-cycle effects of a process, activity, or product will be beneficial in evaluating potential award recipients, but the methodologies and data to conduct comprehensive analyses are not well-developed at this time.*

Innovativeness and Transferability

Since a major objective of an awards program is to encourage groups to launch bold new source reduction initiatives, the innovativeness of various activities should be an additional criterion used. Publicizing innovations in source reduction will result in a more valuable exchange of information than publicizing techniques that are well known. A related criterion that should be considered is the transferability of a product or activity. An activity with little or no relevance to other firms or organizations might in some cases be looked at less favorably than activities with a great deal of relevance. Substituting water for solvents or eliminating unnecessary layers of packaging are examples of innovations that are transferable across a wide range of industries. An organization's willingness to share award-winning design technologies or procedures might be viewed positively by the awards panel.

Economic Benefits

An additional general criterion that should be considered (applying primarily to businesses) is the economic benefit associated with particular source reduction activities. If an activity receiving recognition reduces costs significantly, it will be more likely to motivate others to adopt similar procedures. While economic benefits should be taken into account, innovative source reduction activities should not be precluded from recognition if economic benefits are not immediately realized.

Commitment to Environmental Protection

A final criterion that might be taken into account is some measure of an organization's commitment to environmental protection. While attaching awards to specific products and/or activities diminishes the importance of this criterion, it is desirable to recognize organizations that have demonstrated commitment to environmental quality. Moreover, the awards program may lose credibility if a recipient of a source reduction award is deemed to have a poor environmental record

4. Make the Program Visible

Every effort should be made to ensure maximum publicity and prestige for an awards program. Ways to promote visibility could include launching an aggressive public relations campaign, selecting prestigious individuals to serve on the judging panel, and having the president, the EPA administrator, or some other high-ranking official present the awards at a special ceremony. Information on award-winning activities should be widely disseminated. For awards associated with a specific product, manufacturers could be allowed to publicize their award in advertising for that product. Manufacturers also could be allowed to display a special label on award-winning products.

5. Administer Program Efficiently

Eligible organizations should be encouraged to apply for awards. To help finance the program, nominal application fees might be assessed to corporate applicants. Large businesses could be charged more than small ones.

Various state waste minimization award programs have established independent panels of experts, consisting of representatives from industry, government, academia and environmental groups, to select award winners. A similar panel of about 12 to 20 experts should be formed for a national awards program. Such a panel, comprised as a special national commission, could create enhanced credibility, visibility, and prestige for the awards. Furthermore, having input from various groups in the decision-making process could lead to better consideration of all of the relevant

*See chapter 1 for a discussion of Life-Cycle Assessments.

criteria.

Creating an independent panel of experts to determine award winners could enhance the success of a source reduction awards program. However, a relatively small panel of experts will need assistance in screening potential award-winning activities. In determining award winners for highly technical activities such as production changes and new product designs, the Steering Committee recommends including another decision-making body in the award process. The national panel of experts outlined above would establish criteria for the award and review finalists, while an independent testing organization would review whether applicants meet the technical standards established by the panel. Another option is to coordinate a national source reduction awards program with state and/or regional award programs. In this case, the panel of experts for the national program could review and select award winners from among the winners of the state or regional programs.

LABELING CONSUMER PRODUCTS

Recent market research has shown that many consumers express a high degree of concern over the environmental effects of products and are interested in purchasing products that are environmentally compatible. Labeling consumer products for their environmental attributes is one method that may be used to inform consumers and stimulate MSW source reduction. For more than 10 years, Germany has run a program that awards a "Blue Angel" logo to products identified as better for the environment. Canada initiated its own "Environmental Choice" program about two years ago. In the United States, several states and private organizations have recently begun to formulate environmental labeling programs, and legislative initiatives are under way in Congress.

Environmental labeling programs may be grouped into two types of approaches—"environmentally preferred" or "standards setting." *Environmentally preferred labeling*, as practiced in West Germany and Canada and now in the United States, is voluntary (manufacturers can choose whether to apply), positive (only relatively superior products are distinguished), and based on specific technical criteria.* These criteria can go beyond MSW management concerns and include pollutants generated and energy used during materials extraction, manufacturing, distribution, and use. Logos or seals are awarded to products that are determined to be relatively environmentally sound within a selected product category.

A *standards setting* approach, in contrast, controls the use of certain terms (such as "recyclable" or "recycled content") by defining them and establishing conditions for their use in advertising and product labeling. A standards-setting approach can also be used to control the use of symbols such as the three-arrow recycling logo. If implemented by a government agency through regulations, the standards can be mandatory. Standards developed by industry associations and independent testing organizations can also be implemented on a voluntary basis.†

The following sections explore a number of issues relating to both approaches and provide conclusions reached by the Steering Committee.

Environmentally Preferred Programs

A program to label environmentally preferred products offers the following *advantages*:

- It can harness consumer enthusiasm for environmentally sound products by enabling consumers to identify and purchase preferable products.
- It can reduce confusion and misleading claims by providing a process for independent verification of the environmental superiority of products that receive a seal.
- It can stimulate manufacturers to redesign products and production processes.
- It can enhance general awareness of environmental concerns.

There is evidence from market surveys that many consumers are eager to buy environmentally preferable products. At this point, however, consumers have little access to independent sources of information about the environmental effects associated with products. In addition, products may have different beneficial attributes (for example, minimal packaging or recycled packaging materials) or combinations of beneficial and detrimental attributes (for example, phosphate-containing detergents in packaging that has been lightweighted). Even highly motivated consumers may find it difficult to sort through environmental claims to make the most appropriate choice. An environmentally preferred labeling program can perform the function of evaluating and verifying a variety of environmental effects, thereby providing a basis for consumer choices.

*Green Seal, Inc., based in Washington, D.C., has initiated a national labeling program in the United States.

†An example of this type is the California-based Green Cross program, which operates a voluntary standards-setting program for participating retailers.

To the extent that consumers respond to products bearing an environmentally preferred seal, manufacturers will be rewarded for offering those products and encouraged to develop new products that are better for the environment.

An environmentally preferred labeling program accompanied by educational efforts and public criteria can also be helpful in raising awareness of environmental issues relating to products.

There are several potential *constraints* to environmentally preferred labeling that should be addressed in the design of any program of this type:

- It may be difficult to determine environmental preferability for large numbers of products.
- Labeling has the potential to be misleading to the consumer.
- There are potential legal barriers.
- An environmentally preferred labeling program does not comprehensively control the misuse of claims about environmental benefits.

For many products, it will be challenging to develop and apply comprehensive criteria as the basis for judgments as to their environmental soundness. Some products are clearly preferable (a reusable and durable shopping bag, for example, is better than a paper or plastic disposable bag). However, many products are more complex to analyze. If a product is awarded a label for only one positive attribute, other more serious negative attributes may be ignored.

It is often difficult to evaluate products that have competing negative impacts. For example, powdered laundry detergents may contain phosphates, but phosphate-free liquid detergents may release volatile organic compounds. Methodologies and data are often lacking to carry out life-cycle assessments as a basis for evaluating competing impacts. In the early stages of an environmentally preferred product labeling program, product categories may be chosen because of the ease of development of criteria. If these same categories are relatively unimportant in the market or have relatively insignificant environmental impacts, however, program effectiveness will be constrained.

There is some potential for labels to be misleading to consumers rather than informative. Consumers may believe that products with labels have an absolute environmental compatibility that could lead to misuse of products. For example, consumers might disregard instructions for proper use and disposal of products such as re-refined motor oil or low-solvent paint if they were labeled as environmentally preferred. Depending on how boundaries for a given category are defined, the most environmentally sound products might be discriminated against in the selection process. For example, awarding a label to recycled paper towels might ignore the option of reusable cloth towels.

There may be legal barriers to implementing a labeling program. Although litigation has not been a significant factor in existing environmentally preferred labeling programs, litigation is more prevalent in the United States than in Germany or Canada. Manufacturers may bring a lawsuit if their products are denied the label. There are also questions of liability in the event that a labeled product turns out to be dangerous or defective.

Environmentally preferred labeling only provides information about the products that are labeled. Until a labeling program becomes well-established and applies to many products, the program will not necessarily prevent confusing or misleading claims of environmental superiority in product categories for which there is no label.

Recommendations

The Steering Committee generally supports the concept of environmentally preferred labeling but does not endorse any particular labeling program or make specific recommendations for creation of a specific type of program. Such recommendations are considered beyond the scope of this project. Nevertheless, because of the potential contributions--as well as drawbacks—from environmentally preferred labeling programs, the committee has identified eight conditions under which a program could be most beneficial (figure 3.1).

1. A National Program

The Steering Committee believes that product labeling should be done at the national level, preferably through a single program. A single national program will reduce consumer confusion and reduce the burden on companies of applying for multiple labels.

2. Select Product Categories with Care

An environmentally preferred label typically provides relative—not absolute—information about products within a logically designated product category (for example, paper products, detergents, light bulbs). The choice of categories, therefore, can determine to what extent superior environmental attributes can be identified. For example, fuel efficiency might be used as the basis for awarding a label within the category of automobiles. Such a label would not inform con-

Figure 3.1
Conditions for a Successful Labeling Program

1. The program is national in scope.
2. Product categories are selected to ensure that labels accurately reflect superior performance.
3. Criteria for awarding the label are based on considerations of environmental impacts through the life of a product.
4. Criteria are set so as to stimulate technological innovation.
5. Criteria are made public.
6. Criteria are periodically reevaluated and strengthened.
7. Consumers are educated about the meaning and limitations of the label.
8. Government and industry are consulted and included in the program.

sumers that other products (such as bicycles) can fulfill a similar function (transportation) with even less environmental impact than fuel-efficient cars.

Different programs have approached the definition of categories in different ways. In the German Blue Angel program, for example, low-solvent paints and varnishes are eligible for the logo, while latex paints that contain no organic solvents are not, since they cannot always be used for the same purposes. The Canadian program, however, has differentiated paint categories that are "solvent-based" and "water-based" to avoid this problem. The Steering Committee believes that product categories must be defined to include the most preferred choices.

3. Base Criteria on Life-Cycle Considerations

In determining product criteria and product categories for labeling, total environmental impacts of products must be taken into account to the extent feasible. Environmental impacts emerge from each stage of a product's life cycle, including materials extraction, manufacturing, distribution, use, and disposal and can affect all media (land, water, and air).

It will be challenging to conduct rigorous life-cycle assessments for all products as part of a labeling program. Although some comparisons are relatively easy to make based on inventories, data and methodologies are not widely available to make these estimates for every product in each stage of its life cycle.* It is especially difficult to estimate and compare environmental effects associated with different products. Nevertheless, to be confident that products are not awarded for one positive attribute while more serious negative attributes are ignored, some consideration must be given in a labeling program to the impacts of a product throughout its life cycle. Along with the product itself, the environmental impacts of a product's packaging must also be considered.

In practice, life-cycle considerations often might be simplified by the fact that products within a category tend to differ in only a few respects. For example, for a given grade of motor oil, the relevant life-cycle variable might be its recycled content.

Research to improve methods for life-cycle assessments should be initiated and pursued as soon as possible. Development of better methods can support the expansion of a labeling program.

4. Set Criteria so as to Stimulate Technological Innovation

Within each category, there may be questions concerning the stringency of criteria. Should a program reward relatively few products that are very superior or develop criteria that will allow more products to qualify? This issue should be addressed on a case-by-case basis to ensure that the label produces environmental benefits. At least two approaches are possible. The "best-in-class" approach would single out a product that clearly is superior to others based on some obvious break point in environmental performance. If there are multiple or ambiguous breaks, however, a "threshold" level of performance could be established. Standards should be set high enough to stimulate technological innovation.

5. Reevaluate and Strengthen Criteria Periodically

Within each product category, criteria should be reviewed periodically and strengthened to provide an incentive for continued innovation. The length of time between reviews of criteria could vary depending on the product category. An average length of time might be three years.

*See chapter 1 and appendix A for life-cycle assessments.

6. Make Criteria Available to the Public

Any set of criteria will necessarily involve judgments and assumptions. For a labeling program to have broad credibility, the decision-making criteria and process should be transparent, and there should be opportunity for public review and comment.

7. Educate Consumers about the Meaning of the Label

Labels can be effective tools for educating and empowering consumers, and for encouraging more informed decisions. But, without efforts to educate the consumer about the significance and limitations of labels, labels can confuse and mislead the consumer. Consumers must recognize that awarding a label to a product does not signify that the product is environmentally harmless; rather, it is less harmful than the alternatives. All products result in some waste and utilize raw materials and energy to some extent and must be managed properly when they enter the waste stream.

8. Involve Government in the Program

The Steering Committee believes that, while there are advantages and disadvantages to involving the federal government in an environmentally preferred labeling program, the federal government should be involved to some degree if the program is to be successful. Possible roles for government include the following:

- *An independent non-profit organization administers the program, and the federal government plays an informal role.* The program would use government data and studies, and might use government employees to serve on panels, to the extent allowed by law.
- *The federal government charters and sanctions the program, but the program is administered by an independent nonprofit organization.* The federal government would set the structure and operating procedures of the organization, appoint the commission members, and provide the initial start-up funds. Government employees might provide staff support. Commission members could include representatives from various sectors such as environmental groups, industry, consumer groups, and academia.
- *The program is staffed, funded and administered by a federal government agency.* Such a program would most likely be housed in EPA with participation from other parts of the government, although the Department of Commerce is another option. The program could be supplemented with an advisory board of representatives from industry, environmental organizations, consumer groups and academics, as it is in Germany. Or a board with representatives from various interests could run the program, as is the case in Canada.

An advantage of a program with relatively little government involvement, such as option "a" above, is that such a program could be put in place faster. Securing formal government involvement can be time consuming and may require legislation. In addition, some members of the public may have more faith in a program relatively independent from government.

A program with a larger government role could be advantageous in that manufacturers may be more willing to participate in a program that has government's official sanction. Moreover, establishing a national program will eliminate confusion created by multiple state and independent programs. Initial start-up funds provided by government might enable the program to analyze complex products earlier than would otherwise be possible (though also subjecting the program to the uncertainties of fluctuations in budget allocations). Finally, a government-sanctioned program might be more credible to some potential users than one with no government involvement.

Any labeling program will require substantial resources that must be considered in light of the uncertain environmental benefits. Even if sufficient operating funds for a labeling program eventually could be obtained through application and logo fees, the possibility remains that government resources might be better utilized elsewhere.

9. Involve Industry in the Program

The Steering Committee believes that industry must also be involved in some manner in an environmentally preferred labeling program as well, although precautions must be taken to protect against conflicts of interest. The general advantages of industry involvement include:

- Access to industry's considerable (and often unique) knowledge.
- Manufacturers are more likely to apply for a label if they feel comfortable with the process by which winners are selected.
- Industry support is important for widespread publicity and acceptance of a national labeling program.
- A board that includes representatives of a spectrum of interests, including industry, will have the benefits of exchanges among participants.

Representatives of industry could be involved in a labeling program in several ways, including serving on a board of directors or review panel, serving on a special corporate advisory board, or providing informal consultation.

While industry participation is needed, care must be taken to avoid real and perceived conflicts of interest. These can arise when manufacturers are involved in choosing product categories and in choosing criteria for awarding labels. To minimize conflicts of interest, industry representatives involved in the program should not participate in deliberations involving product types manufactured by their firms or their competitors.

Standard-Setting Labeling Programs

The Steering Committee supports the development of national (or at least regional) standards and guidelines for terms used on product labels and in advertising for "green marketing" (for example, "recycled," "recyclable," "biodegradable"). A standards-setting approach to product labeling has several advantages and disadvantages.

A program to define and control the use of terms related to environmental compatibility would have several *advantages*, in addition to the advantages listed earlier:

- It would provide consumer protection by reducing the confusion caused by inconsistent or inappropriate use of terms in product labeling and advertising.
- It would protect manufacturers against accusations of misleading labeling and advertising.
- Depending on the terms subject to standards, it could provide useful information to consumers on managing items when they are discarded.

With the rise in consumer interest in environmentally sound products has come an increase in manufacturers claims about environmental attributes of their products in advertising and on product labels. Without standards, terms can be used inconsistently. In the absence of agreed-upon definitions, it is more difficult to challenge inaccurate claims or to defend against charges of inaccuracy. The development of standards for terminology used to describe environmental attributes of products could lend consistency and credibility to labeling intended to inform consumer choices. Appropriate standards for product labeling could also inform consumers about how to manage items when they enter the waste stream--for example, by identifying items that should be recycled.

Standard-setting labeling programs have at least two potential *constraints*:

- Accurate labeling will not necessarily help consumers choose the most environmentally sound products.
- It may be difficult to arrive at uniform, agreed-upon standards.

Standards will improve consumer confidence in advertising claims. However, labels and advertising may not provide consumers with all the information needed to make good choices. Several recycling terms that have received considerable attention illustrate the types of challenges in standard setting that will no doubt apply to source-reduction labels as they unfold. For example, a standard might allow writing paper to be labeled "recycled" if it had 60 percent postconsumer content. A consumer would not necessarily know that 100 percent recycled content paper was available and would serve the same purpose. There could be confusion between controlled terms, as well. For example, if a consumer has to choose between products labeled "recycled content," "recyclable," or "biodegradable," what is the preferred choice?

On the second point, the definition of "recyclable" may be particularly difficult. Is recyclability a technical determination based on the existence of technologies to recover materials? Is it an operational question that depends on the availability of local recycling capacity for a material? To what extent do recovered materials need to be used as a feedstock in order for a material to be considered "recyclable?" While operational considerations are controlling in determining whether a material is actually recycled versus disposed, documenting the local availability of recycling capability and labeling products accordingly may be burdensome to manufacturers and regulators.

Defining terms and setting standards relating to source reduction might also be challenging. Standards would probably have to be set by product categories. Within specific categories, terms that might be defined include "reduced packaging," "less toxic," or "more durable."

As is the case with environmentally preferred labeling, standards would be most useful if set at the national level.

Concluding Recommendations

Both environmentally preferred and standard-setting approaches to product labeling provide a means of harnessing consumer enthusiasm for "green" products and verifying claims of environmental superiority.

Each approach addresses somewhat different needs, however, and has different strengths and weaknesses. Nevertheless, the Steering Committee believes that the two approaches may be compatible and even complementary. In the German and Canadian programs, labels are awarded with a brief explanation of why--for example, "made with 100% unbleached paper." Terms for which standards have been set could be incorporated into the label. An environmentally preferred label could thus serve as a complement to a standards-based approach (and vice versa), providing additional information to the consumer. If both approaches are pursued, however, they should be carefully integrated to avoid a confusing proliferation of environment-related messages.

RESEARCH NEEDS AND OTHER RECOMMENDATIONS

In the course of developing the other sections of this report, the Steering Committee identified various priorities for research that should be undertaken if source reduction is to reach its potential. This research can and should be performed by government, business, academia, environmental groups, and others interested in accomplishing source reduction. The committee recommends that research be undertaken to accomplish the following:

- **Develop better and more consistent data on the amounts and kinds of municipal solid waste generated at the federal and international level.** As discussed in chapters 1 and 2, a variety of data deficiencies hinder the development of effective and appropriate source reduction policies and programs. Important gaps in data need to be filled, such as identifying toxic constituents in the waste stream and the amounts and kinds of wastes generated by various sectors (for example, residential, commercial, etc.). EPA and state governments should work together to develop standardized terms and methods for waste composition analyses. More consistent data will allow decision makers to make meaningful comparisons regarding waste generation at the local, state, national and international levels. EPA should then disseminate this data through one of the existing information clearinghouses, so that it can be used in decision making across the nation.
- **Develop methods to measure source reduction.** As discussed in the section on goals in chapter 1, measurement techniques are needed to help set quantifiable source reduction goals, to track progress, and to evaluate the effectiveness of strategies. Measuring source reduction is challenging not only because it is a relatively new enterprise but because of the inherent difficulty of identifying *avoided* waste. Research needs include methods to develop: 1) appropriate baselines for measurement, 2) units of measurement for various types of products, processes, and activities, 3) tools to evaluate trade-offs between toxicity and amounts of waste, 4) ways to communicate source reduction performance effectively, and 5) methods to aggregate source reduction estimates at various levels of interest (for example, by industry, sector, region, nation).
- **Comprehend the sociological/behavioral interactions responsible for waste generation patterns.** As discussed in chapter 1, source reduction strategies can be more effectively designed and implemented with improved understanding of the factors that underlie waste generation patterns and their changes over time. For example, to what extent is waste generation a function of economic factors such as disposable income and the relative prices of goods and services including disposal costs? How important is lack of consumer information or "lifestyle" factors such as preferences for convenience products? How can source reduction strategies take such factors into account?
- **Gain a clearer understanding of the difference between per capita generation rates of municipal solid waste in the United States and other industrialized countries.** Comparisons often are drawn between per capita waste generation in the United States and other industrialized countries, where the rates are generally lower. Such comparisons usually are tenuous because of inconsistencies in the way in which municipal waste is defined and measured. Moreover, some of these differences may be explained by structural factors, such as climate and degree of urbanization, which might be taken as fixed. Others are due to differences in consumption patterns, which can reflect such factors as market prices and lifestyle choices. Research, along the lines identified above, is needed to understand how these various factors interact to affect waste generation rates and to identify opportunities for influencing lifestyle choices in the United States. Such information is needed to evaluate whether or not particular programs can or should be adopted.

- **Identify potential source reduction opportunities by sector.** One of the recommendations in the goals section of chapter 1 is that, after doing the necessary research, federal, state, and local governments, manufacturers, and others should set goals for source reduction. This would entail analyzing individual economic sectors (for example, residential, commercial, etc.) to identify feasible opportunities for source reduction. It would be similar to the process that many of these same entities went through to set recycling goals.
- **Evaluate how strategies for source reduction work in practice.** Tool 4.B in chapter 2 describes various strategies that could be used to achieve source reduction. Some of these have been tried, while others have never been implemented. For those that have been adopted, there is often insufficient analysis to determine how successful they have been. Decision makers need better guidance on when such strategies will and will not work, and what results can be expected.
- **Assess whether it is possible to achieve source reduction through economic incentives or disincentives or to reach consensus on what types of products should not be made.** Tool 4.B of the chapter 2 framework assumes that, if they are effective, economic incentives that incorporate the environmental costs of a good or service into market prices (that is, internalizing all externalities) are always preferable to bans or similar "command-and-control" measures. More research is needed to determine whether mechanisms such as full-cost accounting can achieve satisfactory results. If there are situations where incentives are inadequate, can U.S. society avoid bans and similar measures by reaching a consensus on what types of products simply should not be produced?
- **Develop a handbook for carrying out waste audits to identify source reduction opportunities in municipal solid waste.** As described in chapter 2, waste audits can often reveal changes in operating and procurement practices that can reduce waste generation. Various manuals and guidebooks have been developed to help organizations systematically audit their waste stream. However, guidebooks on auditing municipal waste have generally focused on recycling, and those concentrating on source reduction have been limited to hazardous waste. There is a need for handbooks focused on source reduction of municipal solid waste. Such pragmatic guidance would help manufacturing, commercial and office facilities; schools; hospitals; and other generators to reduce the amount of waste they create.
- **Perform market research on consumer attitudes and perceptions related to source reduction.** As discussed in chapter 2, the probability of successfully implementing a source reduction policy depends on understanding consumer behavior. More information is needed on consmers' understanding of source reduction, what their attitudes toward it are, and how their behavior matches their attitudes. Individual firms may perform such market research, but the information is generally not available in the public sector.
- **Study the economic impacts of using reusable products instead of disposable ones.** Because reusable products must generally be cleaned or otherwise serviced, switching from disposable to reusable products will cause a shift in the type and localition of economic activity associated with the product. That is, disposables creates one set of economic activities (due to raw materials extraction, production, etc.), while reusables create a different set (production of the cleaning supplies, labor to do the cleaning, etc.). There is also a change in the location of the activity, since disposables goods are often produced far from where they are used, while reusables are generally serviced in the same locality the product is used in. As discussed in chapter 2, economic impacts play an important role in determining whether a policy will be implemented. Further study should be done on the transfer of economic benefits resulting from using reusable versus disposable products.
- **Clarify and delineate the trade-offs between recycling and source reduction.** As discussed in Tool 3.A of chapter 2, source reduction and recycling may sometimes be in conflict. A change in materials or design to make a product smaller, lighter, or longer-lasting may make it more difficult to recycle. A better understanding of these trade-offs will result in the optimal policies being chosen.
- **Perform the research necessary to extend the Life Cycle Assessment process to include effects and changes needed.** As described in chapter 1 and appendix A, a Life-Cycle Assessments has three componensts: an inventory, analysis of potential environmental effects, and analysis of the changes needed. Because studies to date have focused on the initial inventory phase, more re-

search must be done on the latter two components. Issues to be resolved include the extent to which common data bases and methodologies can—and should—be developed, whether relative weights can be assigned to various elements of the study, and which types of effects should be included in the analysis.

- **Undertake research on the effect of labeling on consumers and manufacturers.** There currently is a high degree of interest in labeling consumer products for their environmental attributes (see previous section in this chapter). Yet there is insufficient understanding of how labels affect the behavior of consumers and manufacturers. Two aspects require analysis. The first is to understand what labeling can accomplish and what its limitations are. This information is necessary when trying to decide what type of source reduction strategy to implement (as in Tool 4.B of chapter 2). If the decision is made to implement a labeling program, then information is needed on the second aspect—what types of labeling are most effective. This includes how much information to present, how it should be presented, etc.
- **Develop realistic alternatives to toxics in products.** One of the goals of source reduction is to reduce risk by reducing the toxicity of solid waste. This can be accomplished by eliminating the use of toxic materials in a product or, where elimination is not technically feasible, substituting a less-toxic material. Further research needs to be done to develop alternatives that reduce or eliminate toxicity, while retaining the necessary performance characteristics.
- **Evaluate the feasibility of a program for solid waste source reduction similar to the National Energy Reduction Plan in the early 1970s.** During the energy crisis of the 1970s, the nation engaged in a planning process to reduce energy use by all sectors of the economy. The amount and toxicity of waste generated in this country is a critical problem today. An evaluation should be undertaken of the effectiveness of the national energy planning process, *and lessons learned in energy reduction, to help determine whether a similar plan would be appropriate to reduce solid waste generation.*

Appendix A
Product-Life Assessments: Policy Issues and Implications

Contents

This appendix summarizes a forum sponsored by World Wildlife Fund & The Conservation Foundation and the Municipal Solid Waste Program, U.S. Environmental Protection Agency, on May 14, 1990, in Washington, D.C.

Preface

The forum described in this paper was convened in May 1990 by World Wildlife Fund & The Conservation Foundation at the suggestion of the Steering Committee for the Strategies for Source Reduction project. Under a grant from the Municipal Solid Waste Program of the U.S. Environmental Protection Agency, this research project was designed to explore how changes in the design and use of products could help reduce the volume and toxicity of the municipal waste stream. Discussion by a panel representing diverse perspectives on developing and using product life assessments was viewed as a useful way to explore how information from these studies might be used for such purposes as evaluating source reduction methods and opportunities.

The attached summary highlights major issues and points of discussion. It is not a transcript or a detailed set of minutes of all discussion that occurred. Moreover, panel discussion does not necessarily represent the views of the Steering Committee.

Panelists

Gary Davis, Senior Fellow, Energy, Environment and Resources Center, University of Tennessee
Norman Dean, Executive Director, Green Seal, Inc.
Richard Denison, Senior Scientist, Environmental Defense Fund
Mike Flynn, Chief, Criteria and Assessment Group, Municipal Solid Waste Program, U.S. Environmental Protection Agency
Diana Gale, Director, Seattle Solid Waste Utility
Robert Hunt, Vice President, Franklin Associates, Ltd.
Celeste C. Kuta, Toxicologist, The Procter & Gamble Company
Paul Nouwen, Project Leader, Ministry of Housing, Physical Planning, and Environment, The Netherlands
Beth Quay, Manager, Recycling Planning and Programs, Coca-Cola U.S.A.
John Schall, Co-Director, Solid Waste Planning Group, Tellus Institute
Bruce Vigon, Senior Research Scientist, Battelle
Jeanne Wirka, Policy Analyst, Environmental Action Foundation

World Wildlife Fund & The Conservation Foundation Staff

Gail Bingham (facilitator)
Christine Ervin

Executive Summary

Private and public sector demand for product-life assessments has surged over the past year or so in tandem with renewed concern over environmental quality. Sometimes referred to as "life-cycle" analyses in the literature, product-life assessments identify energy, resource, and environmental characteristics associated with the manufacture, use, and disposal of a product. Information from such studies largely has been used by manufacturers to evaluate product alternatives and processes. Recently, however, there has been marked interest in expanding their use for purposes such as consumer marketing, product labeling programs, and government policies affecting the use of certain product materials.

This expanding array of applications brings product-life assessments into the public arena as never before. In turn, there is a need to evaluate many of the policy implications associated with their use and to revisit underlying technical and methodological issues in that context. For example, should studies include full assessments of health and safety risks to consumers and workers? What are appropriate uses of product life assessments and what are potential abuses? How can such complex information be presented to the public in a manner that is both understandable and accurate? These and other policy issues were the subject of a one-day forum convened by World Wildlife Fund & The Conservation Foundation on May 14, 1990.

FORUM OBJECTIVES AND STRUCTURE

By focusing on issue identification rather than resolution, the forum's objectives were to: 1) improve understanding of the potential contributions and limitations of product-life assessments, 2) identify areas of the greatest uncertainty and potential conflict, and 3) contribute to other efforts that are needed to work through various policy and technical issues.

In prepared presentations, representatives from the firms of Franklin Associates, Ltd., Battelle, and Tellus Institute emphasized methodological approaches and challenges from the developer's perspective. Representatives from the U.S. Environmental Protection Agency, the Procter & Gamble Company, and the Environmental Defense Fund emphasized needs and challenges from the user's perspective. A panelist from the Ministry of Housing, Physical Planning and Environment in the Netherlands broadened the discussion further to consider European trends and approaches for dealing with challenges. Other panelists represented the Environmental Action Foundation, Coca-Cola, Green Seal, Inc., Seattle Solid Waste Utility, and the University of Tennessee.

MAJOR ISSUES

Panelists identified numerous issues for further exploration—several of which are highlighted below.

Nature and Scope of Product-Life Assessments

- There may not be one single instrument called a "product-life assessment." Most panelists believed that the nature and scope of studies will vary depending on whether the information is to be used for private sector decisionmaking or for various levels of public policy and intervention. In general, panelists tended to believe that overall comprehensiveness if most critical when the use is for policies that restrict (for example, bans, taxes) or promote the use of certain products or materials.
- Panelists had different views on whether or not risk assessment should be included in the scope of product-life studies and, if so, to what degree. Some believed risk assessment to be vital for comprehensive evaluation and interpretation—particularly if used for public purposes. Others believed that government laws and regulations already addressed threshold levels of risk and/or that risk assessment should be addressed in other analyses; inclusion in product-life assessments seemed unmanageable. Still others believed that varying levels of risk analysis can be included in what is called a product-life assessment depending on such factors as available data and probability of high risk.

Potential and Appropriate Uses of Information

- Product-life assessments are perhaps most useful for specific product comparisons (for example, evaluating alternative materials for specific types of products) rather than determinations on generic materials (such as, plastic or paper). Some panelists believed, however, that needs exist for information that generalizes beyond specific product uses.
- Studies also were deemed useful for identifying

points at which changes, such as modifications in technical processes, could yield the greatest environmental improvement.

Analytic Methods and Data

- Panelists differed concerning the desirability and/or feasibility of explicitly weighting pollutant loadings to reflect their differential environmental impacts. Such weighting would supplement or replace methods that present quantities of individual pollutants or assign an implicit one-to-one weight to individual pollutants that are aggregated in some manner.
- Needs and opportunities may exist for developing common data sources and methodologies for widespread use—particularly for studies to be used in the public sector.
- There are both keen interest in developing "streamlined" product-life assessments to cope with the demand for timely information and concern over how that would best be accomplished.
- Trade-offs exist between using proprietary data and data that can be released to the public. Site-specific and detailed industry data often are released to consultants under confidentiality agreements. Lack of public access to such data, however, inhibits public review and the scrutiny some feel are needed to understand and verify study results.

Communicating Results of Product-Life Assessments

- Presenting information to the public in a manner that is both understandable and accurate involves difficult trade-offs. Product-life assessments rarely, if ever, produce black-and-white results, and product comparisons almost always involve trade-offs that are difficult to evaluate.

Product-Life Assessments: Policy Issues and Implications

On May 14, 1990, a diverse panel of 12 people met to discuss product-life assessments. Sometimes known as "life-cycle" or "resource and environmental profile" analyses, these assessments are used to identify resource effects associated with a product over its entire life. A comprehensive study of this type identifies energy use, material inputs, and pollutants generated during a product's life: from extraction and processing of raw materials, to manufacture and transport of a product to the marketplace, and finally, to use and disposal of the product.

While product-life assessments have been conducted in the United States over the past 25 years or so, their scope and application have varied depending on such factors as client interests and available information. Most studies, however, have focused on comparisons of alternative materials for a given product or package use (for example, beverage containers) and have been used primarily by manufacturers for internal decisionmaking.

Recently, however, there has been marked interest in expanding use and product-life assessments for purposes such as consumer marketing, product labeling programs, and government policies affecting the use of certain product materials. This trend largely reflects an unprecedented concern for modifying product manufacture and use to minimize environmental degradation and corresponding trends in marketing for "green consumerism." Private and public sector demand for product-life information has surged. One major consulting firm, for example, commonly performed about one study per year over the past two decades; during 1990, that number increased to more than a dozen product-life assessments in progress. Other firms are moving into the field and the U.S. Environmental Protection Agency (EPA) has launched a major research effort on product-life assessments.

This expanding array of applications brings product-life assessments into the public arena as never before. In turn, there is a need to evaluate many of the policy implications associated with their use and to revisit underlying technical and methodological issues in that context. The one-day forum was held to identify and explore policy issues surrounding the nature and use of product-life assessments and to better understand implications concerning uses of such information, methodological design, and underlying assumptions. Specific forum objectives, therefore, were to: 1) improve understanding of the potential contributions and limitations of product-life assessments, 2) identify areas of the greatest uncertainty and potential conflict, and 3) contribute to other efforts needed to work through various policy and technical issues.

Materials provided to the panel and a group of about 35 observers are presented in attachment B. These include: the agenda with its proposed set of policy issues, list of panel members, a background paper, survey of selected methods and approaches, and a schematic diagram of product-life assessments.

PANELISTS' PRESENTATIONS*

To help set the stage for later discussion, seven panelists described their respective roles in developing and/or using product-life assessments. The synopses below, while attempting to capture the overall thrust of each speaker's presentation, focus on material that was either unique to that particular speaker or not otherwise discussed in later sections. They are in no way representative of all material presented.

Perspectives on Developing the Product-Life Assessment

Some Background

The panelist representing Franklin Associates, Ltd., traced key developments in the evolution of product-life assessments. For example, Coca-Cola generally is given credit for initiating the first such study in 1969 when it contracted with Midwest Research Institute to study energy consumption associated with alternative beverage contains. The second study, conducted for Mobil Corporation in 1970, compared polystyrene foam and molded pulp meat trays, with special attention on pollution effects. In 1974, a study of nine beverage container systems was funded by EPA to compare resource implications of single-use and refillable containers. That public sector effort set a precedent for analytic comprehensiveness and complexity, and helped forge acceptance of methodological ap-

*See attachment B for an overview to these three consulting firms and their methodological approaches.

proaches that had been the subject of considerable controversy among materials manufacturers.* Over the period of 1975-85, however, studies returned to an emphasis on energy use—largely in response to views that pollution control was being addressed by the new environmental legislation of those years.

The Franklin panelist outlined major strengths and weaknesses of product-life assessments based on that firm's experience in conducting nearly 45 studies by the end of 1990. Perceived strengths were that studies:

- go beyond a single plant boundary or a single environmental impact, thereby broadening perspectives of industry clients as well as environmental groups;
- give good guidance for long-term strategic planning, thereby serving as one decisionmaking tool available; and
- use a quantitative approach based on science and engineering, thereby minimizing subjective biases that may affect study results.

On the other hand, studies:

- provide no "right" answer based on fixed scientific principles (that is, an absolute number equivalent to the point at which water boils, etc.), with outcomes instead based on many different assumptions;
- do not easily produce single "bottom line" answers because of various trade-offs among types of effects (for example, varying perspectives on how to compare resource or pollutant effects can influence one's interpretation of the findings);
- are not always comparable because methodologies and assumptions can vary; and
- are plagued by chronic problems in getting accurate data.

The panelist from Battelle, a consulting firm working on product-life assessments both in Europe and more recently in the United States, pointed to areas of common ground with Franklin Associates with respect to methodologies, data needs, and observations on strengths and weaknesses. Major differences were that Battelle focuses more on postconsumer effects and waste management methods, with analysis of such parameters as toxicity and persistence of the waste stream where appropriate. Other observations underscored why studies can vary significantly:

- The client can have a study tailored to its own product and production process rather than one that compares effects of materials in alternative package/product uses such as container types. Such details as sources of fuel, transportation methods, and suppliers can all be made specific to that manufacturer.
- Scope of analysis can vary, such as whether or not the environmental analysis is a profile of pollutant quantities or goes further into evaluations of risk.
- Overall assessments (for example, scores ranging from poor to excellent) can be used by a manufacturer to improve products, to minimize environmental impacts, and/or to minimize costs in meeting selected objectives.

The panelist representing Tellus Institute offered a perspective much influenced by the nature of its ongoing study for various public sector clients, including the Council of State Governments. That study, which focuses on the environmental impacts of the production and disposal of alternative packaging materials, demonstrates additional sources of variation among product-life assessments:

- The study is driven by public policy objectives rather than analyses of specific products per se. Evaluation of resource effects is considered within the context of a broader economic analysis of how actors will respond to alternative policy measures (for example, taxes, bans) that affect amounts and types of materials.
- The study compares "generic" materials, such as glass and paper, across a variety of packaging/product types rather than within the context of specific packages. In addition to listing and weighting quantities of individual pollutants, the study also expresses environmental effects associated with materials in terms of their market externalities—that is, costs not accounted for in private market transactions.
- All data and assumptions are made available for public inspection and critique. This process could allow data to be built up incrementally over time to create what would amount to a product-life counterpart to economic input-output models.

Perspectives on Using the Product-Life Assessment

The Environmental Protection Agency representative explained how product-life information could help fashion recommendations on such specific issues as: 1) choosing among product materials (for example,

*EPA has not reevaluated approaches to product-life assessments since this 1974 study. The agency's new research effort is being undertaken, in part, to address technical information and needs.

recycled versus virgin contents), 2) identifying priorities and methods for reducing toxicity in the wastestream, and 3) carrying out or evaluating product labeling and award programs. On a broader scale, product-life information is needed to carry out EPA's expanded emphasis on reducing the mount and toxicity of pollution in the environment at large. Among other challenges arising from potential uses are:

- the need to determine when less than comprehensive assessments are appropriate and to devise acceptable streamlined methods for such purposes, and
- the need to ensure methodological comparability among studies for policy coordination and integration.

To that end, the panelist described EPA's new research effort to develop guidelines on product-life methodology. The multioffice study, scheduled to conclude at the end of 1991, plans to survey and build on existing methods, will involve input from both technical and policy-oriented groups, and will use peer review methods to ensure quality results.

The panelist from Procter & Gamble drew a distinction between the dual goals of its corporate policy for environmental quality. First and foremost is the company's goal to ensure the safety of employees, consumers, and the environment—determined largely through the use of risk assessment. These assessments focus on risks associated with consumer use of a product and occupational exposure at the workplace. The second goal is to reduce or prevent burdens on the environment—with the aid of product-life assessments. The latter studies typically focus on single changes in product design or a manufacturing process to determine whether or not a new improvement results. Among the various challenges in developing and using studies, the panelist stressed the need for broadly accepted procedures to help:

- drawing boundaries for which components should be included in the analysis (for example, packaging material, label, dyes, etc.);
- defining which wastes should be analyzed (for example, EPA priority pollutants, all emissions, etc.);
- allocating credits for recycling either in the product system under analysis or the secondary product; and
- factoring in considerations of product performance.

The panelist from the Environmental Defense Fund surveyed the wide range of issues affecting design and use of product-life assessments. The need to revisit and/or evaluate these issues was seen to be critical in light of the expanding purposes for which product-life information is intended. Key choices affecting study scope included: 1) types of impacts (ranging from energy and pollution to aesthetic and social considerations such as equity and convenience); 2) types of releases (for example, routine versus accidental); 3) substances to be included (for example, all chemicals or only those for which there are certain types of data); 4) media of release; 5) types of exposure (for example, general population versus occupational exposures); 6) types of health effects (for example, cancer, reproductive); and 7) routes of exposure (for example, direct versus indirect exposure). The panelist also reviewed the contentious history of risk assessment and identified various issues and problems that parallel those in product-life studies.* Among other observations, he pointed to the following lessons that could be drawn:

- the need to develop standardized methodology and assumptions that permit comparability among studies, and
- the need for full disclosure of information and assumptions—deemed to be particularly problematic given the proprietary nature of most studies to date.

A Perspective from the Netherlands

The panelist representing the Dutch Ministry of Housing, Physical Planning, and Environment—an agency that uses rather than develops product-life information—confirmed that many of the issues raised at the forum were identical to those being grappled with overseas. In the case of access to industry data, problems may be even more acute in the Netherlands. Following presentation of a case study on six milk packaging systems, the panelist drew several major conclusions:

- Despite the uncertainties of product-life assessments today, work should continue to improve this instrument. By doing so, much will be learned about the relative strengths of products (in order to retain those good features) as well as their weaknesses (in order to steadily improve those features).
- Emphasis often is placed on technical issues at the expense of equally important economic and social

*See later discussion of risk assessments in the analytic methods section.

aspects. Product-life assessments should involve a broad range of disciplines.

- All parties having a stake in the study outcome should be involved in the process. Such participation increases the level of support for study results and improves the likelihood of success for policy measures needed to respond to those results. Failure to do so leads to endless rounds of competing studies and claims.
- Agreeing on the methodology and responses to potential outcomes also increases the likelihood of cooperation among various interest groups. Assurances that a manufacturer would have a predetermined amount of time to make any needed product improvements, for example, can reduce perceived threats from conducting a study.
- Product-life assessments should be viewed as only one tool to help reach the ultimate goal of achieving sustainable development in terms of energy use, waste management, and environmental quality.

DISCUSSION OF MAJOR ISSUES

Panelists and members of the audience identified a broad range of issues affecting development and use of product-life assessments (see attachment A). The synopses below focus on issues discussed at some length.

Nature and Scope of Product-Life Assessments

The "scope" of a product-life assessment generally refers to: 1) stages of the life cycle (for example, production, disposal), 2) factors evaluated (for example, energy requirements, pollution), and 3) types of analyses (for example, quantities of pollutants, risk assessment). Panelists generally agreed that a study's scope probably should depend on how the information would be used.* A manufacturer's internal assessment used for making a choice between two different packaging materials, for example, could be quite different than one used as the basis for regulatory controls. In general, it was suggested that the overall detail of analysis would likely increase as the level of use in the policy arena or the level of policy intervention increases.

There was no attempt to explicitly delineate a minimum set of components that would be considered fundamental to any product-life assessment. Instead, discussion focused on issues of scope that were relatively controversial, including risk assessment in general, occupational hazards in particular, and whether accidental events such as oil spills should be included.

*See discussion of criteria linking scope and use in next section.

Why Are Certain Types of Analyses Routinely Omitted from Studies?

Consultants were asked why analyses of occupational exposure or accidental releases were not routinely included in studies. Responses varied. Tellus Institute indicated that such analyses *should* be within boundaries of studies theoretically, but that time and funding constraints served as practical barriers to inclusion. The concept of a product-life assessment as a building-block process would be especially important in this regard, because one study would tend to build on another. This would allow incremental broadening of the scope as knowledge and experience evolves. The other consultants felt that measures of occupational exposure, while feasible, would normally be done within the context of risk assessments—not product-life assessments.

With respect to whether or not accidental releases or events should be factored into studies, one consultant indicated that his company attempts to include everything that seems, or proves, to be important. Thus, accidents occurring quite regularly (every year or two) *would* be included but not highly irregular occurrences. Another agreed that a low probability event is unlikely to have any significant impact relative to one unit of output. Further, some questioned the feasibility of factoring in potential impacts from such catastrophic events as Chernobyl or plant explosions.

Should Risk Assessments be Included?

Panelists differed as to whether or not risk assessment should be included in product-life assessments and, if so, to what extent. Selected exchanges throughout the day are summarized below.

A manufacturing representative stressed that safety considerations are the highest priority of corporate environmental policy. Extensive use of risk assessments takes into account populations exposed, the routes of exposure, concentration, and other considerations. The end result is viewed as a black-or-white decision: either a material is safe by a large margin, or it isn't. In contrast, product-life assessments allow a quantitative assessment of burden on the environment—typically focusing on the analysis of a particular process or material change to make sure there is a net environmental improvement with the proposed product change.

The panelist suggested that in an ideal world *all* effects would be evaluated at each stage. This would lead, however, to such questions as: Is neurotoxicity in this worker more important than this species dying off? Product-life answers for this type of apple-and-orange comparison are unrealistic. In short, if one overlays risk assessments with these studies, analysis becomes unmanageable.

Another panelist argued that strong economic incentives exist to make worker safety a top consideration of manufacturers, such as the need to minimize costs for workers compensation, insurance, etc. With that in mind, the need to include risk assessment of worker exposure in product-life assessments might be unnecessary.

A third panelist also agreed that risk assessment is critical for determining safety of the product for consumer use as well as for employees in the workplace. Going further, however, he stated that risk assessment plays a similar role throughout the product life cycle and should not necessarily be confined to one or two stages of the analysis. The panelist drew on experience with risk assessment over the years to suggest lessons for dealing with product-life assessments today. Debate over risk assessment, for example, became contentious and engendered a low level of public trust. Critical in that debate was the artificial distinction between risk assessment (deemed to be pure science) and risk management (considered policy). This bifurcation frequently resulted in the use of risk assessments to defend the status quo rather than as a means to challenge and improve the system. Also, debate focused on specific inputs and outputs—that is, what the analysis could and could not do. By not treating ambiguities upfront—including the areas of greatest uncertainty and policy-related issues—risk assessment was set back in its development. As an example, people assumed 10 years ago that indirect effects (for example, eating contaminated food) were small and that direct routes (for example, air pollution that you inhale) were what counted. Now it is recognized that indirect routes are at least as important and maybe more so. Dealing with such uncertainties at the front end of the process, argued the panelist, might have led to earlier recognition of relative effects.

A panelist responded by agreeing that risk assessments need to be improved, but their routine inclusion in product-life studies would make the whole system unmanageable. For example, how would health and safety considerations figure into a product labeling program? In such a case, the label would be intended to state that the product minimizes environmental impact. For other uses, perhaps a product-life assessment could require analysis of risk.

In conclusion, one individual observed that it all comes down to how one defines "environmental impact." Whether or not impacts should include considerations of health and safety throughout a product's life, and to what extent it should be built into methodology, will need to be evaluated and resolved by policy makers in the future.

What about "Streamlined" Assessments?

Since assessments performed in the past have been very time- and labor-intensive, several panelists expressed the need to determine when streamlined assessments are appropriate and how they should be constructed. This was particularly appealing given that demand for product-life information is likely to increase dramatically and that the nature and scope of studies could vary depending on how the information would be used. Concern was expressed that if we wait to resolve all the complex issues surfacing at the forum, and wait to conduct comprehensive assessments for myriad products, we will never accomplish anything meaningful. Something short of a full product-life assessment seemed prudent and desirable, although there was no discussion of what that might look like exactly.

Staff to one of the consultants stated that public policy simply cannot be applied at the individual product level. Pronouncements on ketchup bottles, mayonnaise jars, and vinegar bottles, etc., are impractical and not very useful. A labeling program that includes only 13 or 397 different product uses, for example, is not meaningful and requires much more comprehensiveness. Instead, product-life assessments need to evaluate uses of generic materials across products to contribute meaningful information for public policy. Another panelist observed that both Germany and Canada had rejected comprehensive product-life assessments in favor of abbreviated versions for their labeling programs and questioned whether or not we had to repeat that discovery process.

Others were concerned about possible consequences of taking shortcuts. One panelist, for example, noted that streamlining poses the substantial risk of making wrong decisions. As an example, he pointed to the fact that study results vary depending on the size of containers analyzed—thus restricting the ability to generalize across sizes. In addition, it was noted that results of product-life assessments can have significant effects on the free market system. Had shortcuts been taken years ago, certain products deemed inferior might have been taken off the market—even though they have now been shown to be superior products.

Uses of Product-Life Assessments

Product-life assessments can produce a wide array of information on energy, resource, and environmental releases associated with a product over its life cycle. The specific uses to which that information is put became a pivotal consideration in most of the issues discussed throughout the forum. Two major distinctions were made: 1) public versus private use of information and 2) consumer information versus regulatory use. Decisions affecting the potential breadth and depth of a study, need for common methodologies, and degree of public disclosure of data and assumptions all were seen to hinge on these distinctions.

What Are the Potential Uses for Product-Life Information?

Panelists first identified the range of potential uses of product-life information through individual presentations and group discussion without attempting to seek agreement on which uses were appropriate or not. As figure A.1 shows, these can be categorized according to whether or not the user is a private manufacturer or public decision maker, although some uses are common to both.

What Are Appropriate Uses of That Information?

Discussion regarding appropriate uses of product-life information occurred at various points throughout the day. In general, panelists agreed that product-life information was most appropriate for at least the following:

- internal decision making by manufacturers;
- comparative purposes in general versus absolute determinations; and
- the specific use in product labeling decisions.

Opinions were more varied, however, as to whether or not assessments could appropriately be used for:

- legislative or regulatory restrictions on materials (taxes, bans);
- evaluating generic materials versus specific products;
- conveying information to the public; and
- promotion of products by manufacturers or trade associations.

Should Product-Life Information be Used for Comparative Purposes Only, and How Does That Affect Its Role in Public Policy?

Panelists discussed whether studies might be used to product *absolute* information on a particular material/product or only in a relative sense to *compare* alternative materials/products. The difference can be important in terms of policy implications. Absolute determinations are more likely to be used in policy measures such as taxes and bans to restrict the use of particular materials. Comparative determinations are interpreted within specified contexts, such as within a product category for labeling purposes or in private decision making. As such, the issue of absolute versus comparative uses closely parallels the issue of whether or not assessments can provide information on *generic* materials (for example, plastic versus paper) as well as *specific product uses* (PET versus glass containers for soft drink beverages).

Should Product-Life Assessments be Used for Policy Intervention?

Policy intervention was viewed to include both intervention by government into the marketplace and intervention by industry into the public policy arena. One private sector representative opposed use of product-life assessments as the basis for bans and taxes because inaccurate data and other uncertainties would lead to faulty generalizations. For example, environmental impacts of containers can vary significantly according to parameters such as size and design. Judgments based on generic materials would be flawed.

Another panelist noted that legislation is enacted in many areas characterized by incomplete information, thereby weakening that argument as the basis for opposing legislative outcomes. Similarly, a consultant suggested that whether or not information should be used for legislative purposes is a moot issue: public policy will and is being developed for materials use around the country. Today that is occurring in a vacuum. The challenge, then, is how to develop studies that are most informative to the policy formation process.

Other perspectives put the issue in a somewhat different light. For example, one panelist maintained that product-life assessments should be used to *inform* public policy, which differs from actually serving as the sole basis for making regulatory decisions. The comprehensiveness of that information will vary depending on its proposed use. Comprehensive analysis is needed for regulatory decisions affecting

Figure A.1
Potential Uses of Product-Life Information*

Manufacturers could use information from product-life assessments in their overall product design and manufacture to:

- compare generic materials;
- evaluate resource effects associated with particular products, including new products;
- compare functionally equivalent products;
- compare different options within a particular process with the objective of minimizing environmental impacts;
- identify processes, ingredients, and systems that are major contributors to environmental impacts;
- supply information for product and procurement audits;
- provide guidance in long-term strategic planning concerning trends in product design and materials;
- help train product designers in the use of environmentally preferred product materials;
- evaluate claims made by other manufacturers;
- enhance market competitiveness; and
- provide information to consumers about the resource characteristics of their products or materials.

Public Decision Makers could use information from product-life assessments to:

- supply information needed for legislative or regulatory policy that restricts use of product materials (for example, bans, taxes, etc.);
- supply information needed to set standards governing product advertising (for example, defining recycled content, etc.);
- gather environmental information;
- identify gaps in knowledge and research priorities;
- help evaluate and differentiate among products for labeling programs;
- provide information to the public about the resource characteristics of products or materials;
- help develop long-term policy regarding overall material use, resource conservation, and the reduction of environmental impacts and risks posed by materials and processes throughout the product life cycle;
- evaluate claims by manufacturers; and
- evaluate resource effects associated with source reduction and alternative waste management techniques.

*This list does *not* differentiate between potential and appropriate uses of information. See discussion in text.

only a few products while much less detail is needed to supply consumers with general information on many products. Finally, the European panelist maintained that studies must be used as the basis for policy in which legislative measures such as taxes and bans could be viewed as the most extreme applications. He cautioned, however, that considerable information is needed for legislative measures addressing generic materials because affected interests have so much at stake in the outcome. Are there useful criteria for linking scope of the product-life assessment with intended use of the information it generates?

Panelists frequently returned to the question of how to relate comprehensiveness of studies and the range of products to be evaluated with intended uses of product-life assessment information. As the following matrix suggests, relatively few products would undergo fully comprehensive assessments as may be required for regulatory restrictions on products or materials. Less detailed analysis that could be performed more easily for a greater number of products might be allowed for purposes of consumer information such as product labeling. This might also be the case when the effects of a material or product throughout its life-cycle are considered relatively benign.

Analytic Methods and Data

At attachment A shows, numerous and complex issues were raised about analytic methods. Those discussed in most detail concerned sources and quality of data, proprietary restrictions on data, common methodologies and guidelines, and weighting of environmental and other effects.

How Do Consultants Obtain Their Data?

All panelists agreed that getting accurate data is a chronic problem with which to contend. Those problems are complicated by proprietary concerns over technical detail.

Franklin Associates relies heavily on data obtained directly from industry, rather than published sources, and attempts to confirm these data to help assure reliability. Where data are particularly scarce, such as for effects of extracting raw materials, staff often construct models for industry review and comment. The panelist for Battelle added that obtaining data from industrial sources does not mean getting it over the phone. On-site plant visits often are necessary to obtain needed information. Each study, and major step therein, requires an evaluation of whether primary data needs to be collected or whether some form of extrapolation will suffice. He also noted that Battelle does not assess the raw materials extraction phase because of the lack of reliable data. Finally, the panelist for Tellus Institute explained that published sources are used for such items as criteria pollutants, whereas more detailed industry data for specific plants are used when available and appropriate.

Another panelist stressed the complexity of data needed for product-life assessments. For example, in the analysis of aluminum beverage cans, aluminum is sourced from various parts of the world from manufacturers that may employ differing technology and processes. In the United States alone, for example, cans used by the Coca-Cola Company are manufactured from sheet material drawn from more than 60 different locations.

What Are the Issues Surrounding Confidentiality and Public Access?

Agreements to maintain the confidentiality of certain data are an important way of doing business for Battelle. To get the best information available, the consulting firm often must act as an intermediary between the manufacturer and its suppliers. Two and even three-way confidentiality agreements encourage provision of more detailed data but also serve as a barrier to public access.

Franklin's extensive data base includes all levels of confidentiality, from wholly public sources (used for publicly funded studies) to components owned substantially by the client. Like Battelle, it is common to have agreements protecting information between clients and suppliers. In short, confidentiality involves a delicate balance in maintaining good relationships with all clients, including those who compete against each other.

The public study being conducted by Tellus Institute uses a mixture of private and public data but will make all background data and assumptions available to the public. The panelist observed that because the study will affect public policy decisions, industry associations and individual companies have cooperated in making the data as accurate as possible. When asked if there were often disparities between private and public data, the panelist said no. Typically discrepancies reflect outdated information such as whether or not data are based on a new production process, etc.

Several panelists pointed out that without access to underlying data and assumptions, study results cannot be fully understood, let alone verified—opening the door to endless disputes. This was deemed particularly important now that study results are being used to influence public attitudes and choices in the market. The Tellus consultant explained the building block concept in product-life assessments: that data bases and methods could be built up incrementally with each new effort to produce an equivalent to the matrices making up economic input-output models. That cannot occur, however, if data and assumptions are not open to scientific inquiry and critique.

Is There a Need for Generic Data Bases?

Some panelists saw the need to develop generic data bases to offset the problems raised with proprietary data. This would allow the public and scientific community to better understand studies and to scrutinize data for accuracy; knowledge would be steadily built up and improved over time. As discussed above, however, others were concerned that loss of confidentiality would result in more reliance on industry averages and, therefore, less precision. Industries would have very little incentive to supply detailed data that would be released to the public. While industry associations might be used to process proprietary data, one panelist expressed disappointment with examples of that process.

Finally, several panelists believed it desirable to

identify circumstances in which it is appropriate to use generic versus site-specific data and, further, to develop generic data bases with guidelines for their appropriate use. The latter was viewed as important especially for smaller firms, which may not be able to afford site-specific data collection. Sensitivity analysis could play an important role in developing guidelines: if significant changes in certain input data do not affect results appreciably, the accuracy of those data will be less important.

What Are the Potential Advantages and Disadvantages of Using Common Methodologies?

Since product-life assessments were first performed in the late 1960s, a certain amount of closure has been reached within the private sector on acceptable approaches and methodology. Nonetheless, practices can vary considerably depending on the study consultant, the nature of the study, and interests of the client for whom a study is prepared as suggested in the review of consultant presentations. The issue of whether or not common methodologies are needed for product-life assessments was viewed largely in response to the merging public nature of study uses—a theme recurring throughout the forum. In other words, most panelists were far less concerned with the need for common methodologies of one sort or another if the study were conducted solely for private decision making.

Some panelists believed that product labeling in particular demanded a common framework to assure equity among affected parties. Such commonality also would make it easier to provide meaningful results to the public. Constantly changing methods and assumptions would exacerbate existing problems in communicating complex information to readers. The EPA representative also believed common guidelines were necessary to ensure comparability among a wide range of studies that the agency needs to conduct.

Others were concerned about possible drawbacks to standardization. For example, one consultant stressed that flexibility to deal with day-to-day developments is important for keeping methods responsive to changing circumstances and new ideas. Another panelist in the public sector was concerned that premature attempts at standardization would tend to stifle innovation. Instead, competition among several parties might be a productive way to thrash out various issues, to experiment, and to test assumptions. This observation led to the exchange below.

An Exchange of Views: Will Lack of Standardization Result in "Fly-by-Night" Studies?

One panelist cautioned that lack of guidelines could precipitate problems in quality control. Unlike the past, various parties now have a substantial staked—either political or financial—in the results of studies communicated to the public. This new context may result in a proliferation of consulting firms producing "fly-by-night" studies. Another panelist, however, contended the opposite result could occur based on what happened when EPA issued guidelines for bioassay studies in toxicity reduction work. Whereas a small number of firms were involved prior to the guidelines, there followed a proliferation of consultants who "set up their fish tanks" and got into the business of bioassay research.

In response, it was noted that such backyard consulting was likely to occur with or without the guidelines. Underlying incentives would drive that trend. The critical issue, then, is how to exert quality control to cope with that proliferation.

An alternative to complete standardization, suggested by several panelists, was to provide general guidelines rather than "cookbooks." Performance criteria could be specified, allowing developers to use discretion and creativity in meeting those criteria. In summary, many panelists believed standardized methodology to be most important for public uses of product-life information.

Should Weights Be Assigned?

Several panelists believed weighting, the relative ranking of effects, to be the single most important and complex issue of product-life assessments. Panelists disagreed as to whether or not relative weights should be assigned to different types of pollutants both within air, water, and solid waste media, and across media. Further, the issue of weighting extends beyond environmental impacts to the comparison of other impacts ranging from energy and resource effects to those affecting such considerations as product performance.

Several panelists questioned the need for ranking since governmental laws and regulations should suffice—at least implicitly—as a relative measure of pollutant impacts. For example, why shouldn't emission standards reflect some level of safety afforded the public?* If not, are product-life assessments going to

*Panelists did not explicitly discuss weighting implications—that is, should the presence of emission standards mean that pollutants emitted at or below their respective standards would be weighted equally.

be used to change emission standards? Given the difficulty of weighting, these panelists felt they needed to work from some baseline assumption such as government standards to make product-life assessments feasible.

Other panelists believed that the unfolding interest in pollution prevention in general, and product-life assessments in particular, reflected public concern regarding the inadequacy of governmental standards. One individual expressed this as a growing frustration with the existing pollution control culture. Others pointed out that regulations are not always comprehensive. For example, mining and recycling wastes largely are exempt from regulatory control. Likewise, many pollutants lack individual standards. The discussion that should be occurring, therefore, is how to evaluate relative environmental effects, rather than identifying their quantities alone.

Can Weights be Assigned?

Regardless of its merits, some panelists questioned whether or not ranking *could* be done since reaching a consensus on the rankings will be difficult, if not impossible. Others believed that, despite the difficulty, there are tools available to achieve consensus, such as Delphi techniques. Attention should focus on when weighting should occur and methods for doing so, which should always be made explicit.

Still another believed that relative rankings of such diverse effects as a pound of carcinogen and a pound of sulfur dioxide would not be adequate. Instead, a quantitative listing of all pollutants would be more informative.*

Who Should Do the Ranking?

One panelist pointed out that ranking will occur in one form or another. If product-life specialists don't do it in at least a quasiscientific manner, politicians will. At least one panelist believed the latter might even be preferable because the issue of relative weights really translates into the much broader realm of social values. Since this realm lacks technical methods, politicians are perhaps the most appropriate ones to rank. To do that, they need information on consequences of their choices.

*Anything other than listing individual pollutants is a form of weighting, be it implicit or explicit. Thus, aggregation into pollutant classes (for example, total hydrocarbons, biological oxygen demand) or into media classes (for example, water discharges) involves weighting.

How Should Weighting Systems be Presented?

While the longer run issue may be how to devise weights, an immediate issue raised by one panelist is how to explain whatever weighting system is used. What caveats need to be presented so that the public and press can understand implications? In one recent study, it was noted, a consultant's report stated that comparison of pollutant impacts for the different products could not be attempted based on the weighting system used—but the client drew different conclusions in public statements.

Panelists pointed out that it is important to know when weighting actually is occurring. Aggregating amounts of individual pollutants (for example, total amount of hydrocarbons or even all air pollutants) implicitly rank everything as equal. Since that represents a weighting system, it should be explained fully for proper interpretation.

What if no weighting system is used—that is, if amounts of each and every pollutant are enumerated? Several panelists pointed out that this alternative has problems also because it doesn't necessarily allow the layperson to understand the information provided: everyone would have to act as their own toxicologist. The question then becomes how to do this intelligently? The fact remains that those who perform assessments will always know more than those to whom information is being conveyed. Simplification tends to lose information in the process. One audience member noted that we might draw analogies from experience with food labeling. Certainly, food labels don't list everything but tend to list items (for example, sodium, cholesterol, etc.) that may affect different types of individuals.

In summary, most if not all panelists agreed that presentations should explicitly address whatever weighting scheme is used.

When Might Weighting be Least Critical?

One panelist suggested that there might be instances in which weighting is not crucial. For example, if all pollutants or other factors decrease as a result of changing from x to y, then aggregation of pollutants may not be critical in drawing conclusions. In a recent study of soft drink containers, to cite one example, energy requirements and pollutant quantities consistently decreased as the size of a container made from the same material decreased.

What Is the Common Denominator to Use for Comparisons?

The panelist for Tellus Institute viewed the common denominator for comparisons to be dollars, consistent with the approach used in Tellus's packaging study. Eventually the assessment of environmental damage, or the cost to clean up, is translated into monetary values. While recognizing this always will be somewhat subjective, no alternatives seemed apparent. Sensitivity analysis, once again, was seen as an important tool to gauge levels of uncertainty in the underlying data. If changing the weighting by 10 percent reverses the order of results, uncertainty in the data becomes important. If you change weighting by a factor of two and results don't change, then uncertainty may not be an issue.

Communicating Results

Effective and accurate portrayal of product-life assessment results was considered a significant challenge by all panelists, particularly given the trend for using study results in the public or consumer-market arena.

What Are the Trade-offs between Effectiveness and Accuracy?

Consultants were asked to describe constraints they face in presenting study findings. One characterized the advice from public relations experts this way: "If there's more than one number in a study, you're dead!" While intended as a slight exaggeration, the maxim aptly conveys how important it is to present information simply if maximum impact is to be achieved—particularly for use with the general media where emphasis is placed on the results alone, the proverbial "bottom line."

The trade-off, of course, is that information may be lost that would allow readers to accurately interpret study results based on the subtleties of various methods and assumptions. In short, most information contained in a product-life assessment is not so black-and-white that it lends itself to simple interpretations and bottom-line numbers.

How, then, can the competing objectives of effectiveness and accuracy be optimized? One panelist working in the public sector noted that she had more faith in the public's ability to absorb complex information, and that one simply had to be careful. For example, study results can be taken to legislative bodies, including well-prepared executive summaries, with explanation of the various implications. Such complex information can be publicly presented in such a way that makes assumptions and potential biases clear and makes trade-offs obvious. Another panelist agreed that it was critical to include all important assumptions and qualifications in the executive summary itself since competing interest groups often used the summary alone without reference to substantive qualifications contained elsewhere in the study.

Nevertheless, the Dutch panelist stressed that misinterpretation of findings is a risk and that the public can and does make choices on the basis of these studies. In the Netherlands, for example, the public has demanded that PVC plastics be banned on the basis of a recent study.

How Do Consultants Control Use of Their Studies?

Consultants were asked how much control they have over the use of their studies once completed. The Franklin panelist explained that a client typically owns the data base, but there are contractual restrictions on how the data can be used. Franklin can also review and comment on followup material developed by a client, but whether or not the clients follow that commentary to the letter is more difficult to ensure. Battelle added that their contracts prohibit use in advertising, but that court action through lawsuits would be their only recourse should that be violated in some way.

"Rock-in-a-Box" and Other Challenges

One manufacturing representative stressed that they use product-life assessments to provide factual data to the public rather than as an advertising tool to "pit" one product against another. Instead of advertising labels per se, it was more meaningful to explain why product concentration and recycling, for example, result in environmental improvements. She also stressed the challenge in educating people to understand that *all* products have impacts and trade-offs. If not—if their only goal were environmental quality and not product performance and other objectives—the perfect detergent would be simple: a rock in a box. Factoring in product performance, therefore, was deemed especially important by industry for uses such as labeling.

Some Guidelines for Public Presentation and Use

Various suggestions were made by panelists and audience members to help minimize problems in communicating results to the public, including some already discussed above. Without seeking to reach agreement on their merits, these included:

- Clearly identify in an executive summary all pertinent assumptions and qualifications that may affect interpretation of findings.
- Present enough data so that readers can draw their own conclusions. Don't present a ready-prepared conclusion alone that readers must accept or reject.
- Consider the possibility that some results should *not* be presented to the public, particularly in cases where there is a high risk of misinterpretation.
- Use focus groups to determine how the public might interpret the information.
- Consider use of peer review mechanisms when product-life information might be used in the public arena.
- Be careful to place findings within a realistic and meaningful context—that is, differentiate between product findings that are truly significant and those that aren't.

Attachment 1
Range of Policy Issues Identified in Forum

Defining the Nature and Scope of a Product-life Assessment

- What *is* a product-life assessment (PLA)? What stages of the life cycle and types of analysis should be includes?
- Should and how does its definition and scope vary according to use? Is there some set of minimum components? Under what circumstances should a complex versus "streamlined" PLA be conducted? What would a streamlined PLA look like to be defensible?
- Should the following types of analyses be included: risk assessments? conservation of resources? economic analyses and valuation? social values such as equity and convenience? product performance? aesthetic impacts?

Potential and Appropriate Uses of Product-Life Information

- What are the various potential uses of PLA information? Which uses are appropriate and inappropriate? When should a PLA *not* be done given level of uncertainty or some other factor?
- How should/will public decision makers apply PLA information? Can information be used as the basis for regulatory restrictions on products, materials, or substances? Are PLAs an adequate tool for measures to improve environmental quality? How does uncertainty affect usefulness for public policy decisions?
- Should PLAs be used for comparisons alone or in an absolute sense? Can they be used to generalize beyond specific products to address materials?

Analytic Methods and Data

- Should environmental impacts versus amounts of pollutants be estimated, given the existence of environmental and safety laws and regulations? How can relative weights be assigned to pollutants within and across media? How can other considerations such as product performance and safety be compared with environmental, energy, and resource effects? Who should determine relative weights?
- Is there a need for common methodologies or guidelines, and where would these be most important? Should generic data bases be developed and for what kinds of information? How can quality of information be assured? Could a measure of data quality be created? What are the constraints of confidential business information, and how can they be addressed?
- How should uncertainty be dealt with and presented? Should public policy preferences for source reduction versus treatment of wastes be incorporated in analysis and how? How much should be spent for increasing detail and accuracy? When do expenditures outweigh benefits?

Communicating Results of Product-Life Assessments

- How can information best be communicated to the public? What are the trade-offs between effectiveness in presentation and accuracy of information? How can levels of uncertainty be conveyed? How can statistics be presented in a useful and meaningful way?
- Should all data and assumptions be made available for public inspection, and what are the challenges and trade-offs in doing so?
- How can assumptions be presented in such a way as to promote clear interpretation?

Attachment 2

Survey of Selected Methods and Approaches of Panel Consultants

(distributed at May 14 Forum)

In preparation for the May 14, 1990 forum on Product Life Assessments (PLAs), a survey questionnaire was completed by several forum participants who conduct such studies. This paper summarizes questionnaire responses by individuals representing the following organizations: Battelle, Franklin Associates, Ltd., and Tellus Institute. (A copy of the questionnaire is included with other materials.)

Material presented in this summary follows individual responses as closely as possible and is subject to review and comment by the consultants. Further, the summary is intended to indicate the various approaches used in their PLAs—based in part on the different purposes for which they have been designed, rather than a precise description of methods and assumptions.

Background to PLA Consultants Surveyed

Battelle is a multi-disciplinary, nonprofit research and development organization based in Columbus, Ohio. Founded in 1929, Battelle employs 7,000 staff worldwide including about 600 staff working in areas directly related to health and environmental issues. PLA work was undertaken in this country as recently as 1988, but Battelle's experience in Europe dates into the 1970s. Battelle's studies are termed "life cycle assessments" in the U.S. and "ecobalances" in Europe. They have been performed exclusively for individual business clients for the following purposes: 1) to identify potential problem materials, 2) to set up recycling or reuse programs, and 3) to compare alternative materials in the same application.

Franklin Associates, Ltd., located in Prairie Village, Kansas, was founded in 1975 as a technical environmental consulting firm specializing in solid waste management. With approximately 28 staff, most of whom are engineers and scientists, Franklin Associates has conducted more PLAs than any other single firm in this country. Franklin staff conducted the first known PLA-type of study at Midwest Research Institute in 1968-69 and have conducted nearly 40 studies since then. Called Resource and Environmental Profile Analyses (REPAs), about 70 percent have been performed for individual businesses; 20 percent for industrial associations, and 10 percent for the federal government. Purposes of such studies include: 1) estimating environmental consequences of alternative materials for long-term planning, 2) legislative policy, 3) marketing, and 4) new product analysis.

Tellus Institute, formerly known as Energy Systems Research Group, Inc., is a multidisciplinary, nonprofit research and consulting firm located in Boston, Massachusetts. Founded in the mid-1970s, the 60-staff organization has expanded its original focus on public utility and energy research to address risk analysis and other environmental issues. The Solid Waste Group initiated its first PLA in January 1989 for the state of New Jersey, the Council of State Governments, and the U.S. Environmental Protection Agency. The project is designed to: 1) evaluate the environmental impacts of production and disposal of alternative packaging material, 2) analyze public policy measures to alter the mix of packaging used, and 3) analyze economic impacts of such measures. Multiple PLAs are included in the packaging project; two PLAs are partially complete and twelve more are underway as of May 1990. The overall purpose of the PLAs is to assist government policy decision-making concerning the use of packaging materials.

Attached Summary Tables

The attached tables provide a side-by-side comparison of responses to the questionnaire.

- Table 1 shows which factors (i.e. energy requirements, material conservation, type of pollutant releases) are considered at which stage of the product-life (i.e. raw material extraction, material processing, product manufacture, distribution, use, and disposal.)
- Table 2 summarizes the types of environmental analysis performed and various methodological approaches.
- Table 3 describes other considerations in the PLA, including how studies address recycling/reuse rates, landfill and incineration effects, and economic analysis.

Table 1
Scope of Product-Life Assessments by Selected Organizations — Types of Considerations

Stages of product life	Energy	Materials conservation	Air emissions	Water effuents	Industrial solid waste	Postconsumer soid waste
Extraction of raw materials	F/T	F	F/T	F/T	F/T	—
Processing of materials	B/FF/T	F	B/F/T	B/F/T	B/F/T	—
Product manufacture	B/F/T	F	B/F/T	B/F/T	B/F/T	—
Distribution / transportation	B/F/T	F	B/F/T	F/T	F/T	—
Product use	F	F	B/F	F	F	B/F
Postconsumer recycling	F	F	F	F	F	B/F/T
Disposal	F/T	F	B/F	B/F	F	B/F/T

Key to letters: Presence of the following letters signifies that organization considers this fact at given stage in product life.

B = Battelle
F = Franklin Associates, Ltd.
T = Tellus Institute

Note: This table has been prepared for review purposes.

Table 2
Considerations in Analysis of Environmental Wastes
(Air / Water / Solid Waste)

PLA analysis considers:	**Battelle**	**Franklin**	**Tellus**
Amount / volume	Yes	Yes	Yes
Toxicity	Yes	No	Yes
Exposure	Only where generic pathway is defined	No	No
Persistence	Via mechanical breakdown / degradability	No	Possible; methods being developed
Mobility	Via surrogate measures (e.g., water solubilty)	No	Possible; methods being developed
Global effects (e.g., climate change, ozone depletion)	Qualitative only	No	No
Risk assessments	Not as part of PLA	No	No
Consumer/worker safety	No	No	No
1. For releases to the environment, what criteria are used to select pollutants to measure?			
a) pollutants covered by federal/state laws and regulations	Yes	Yes	Yes
b) pollutants that exceed some threshold level, regardless of regulatory controls.	Yes	—	—
c) impact of pollutants (e.g., toxicity, etc.)	Most important criterion	—	Yes
d) SARA 313 list of toxic chemicals	—	—	—

Table 2
Considerations in Analysis of Environmental Wastes
(continued)

PLA analysis considers:	Battelle	Franklin	Tellus
2. Are releases assumed to meet current treatment standards?	Not always. Technology mixes are devised for projections. Also, a mix of old and new technologies is used for landfills analysis.	Only if actual emission data are not available.	Only if actual emission data are not available.
3. Is *impact* of individual pollutants estimated?	Try to assess whether concentration may be a problem only where a defined pathway and threshold level exist.	No	Methods under development
4. What about relative impacts within and across media?	—	No	Methods under development to rank relative impacts, especially within media
5. Is analysis primarily quantitative or qualitative?	Mix of qualitative / quantitative depends on product stage and environmental pathway	NA	Quantitative

Table 3
Other Considerations in the PLA

How are the following items considered?	Battelle	Franklin	Tellus
a) Recycling rates of the packaging or material type	"Fatetree" model includes general recycling rates for a material; also, analysis of increased recycled content.	Estimates actual rates and/or looks at various recycling rates from 0 to 100%.	Estimates recycling rates with emphasis on commercial feasibility.
b) Reuse of the packing or material type	Analyzed as possible fate pathway, including transport of material to produce and steps needed to put material back into use.	Estimates actual rates and/or looks at various reuse rates from 0 to 100%.	Will consider reuse of glass.
c) Landfill/incinerator impacts (e.g., leaching, emissions, etc.)	Models based on amount of materials allocated to facilities according to published literature (unless data specific to product are collected)	Energy released and incinerator ash produced. Others more difficult to tie back to individual product.	Not considered.
d) Economic considerations	Estimates trade-offs among environmental / energy considerations, manufacturing costs and other materials performance criteria. Other indirect costs are either implicit (e.g., wastewater treatment during manufacturing) or are not considered.	Not considered.	Disposal impacts and economic impacts of taxes and cans.

Appendix B
Application of the Framework to Household Batteries

The Steering Committee initiated research into several product categories* during the course of developing the Evaluation Framework in chapter 2 in order to test the framework. The research into batteries is presented here to help demonstrate how the framework might be applied. This material does not represent a rigorous analysis of household batteries and is not intended to prescribe solutions. Its only purpose is to illustrate how the framework might be used. Readers may wish to refer back and forth between chapter 2 and this appendix to compare how the questions are asked in the framework and answered in this example.

The paper begins by answering the questions in Tool 1, based on the information that one might have at hand before conducting an in-depth analysis of a product category. There is also an example of how this information might be summarized in matrix form. A sample of the Tool 2 checklist indicates the options that might be checked off at this point in an analysis. This is followed by a sample of how Part A of Tool 3 could be filled out for household batteries.

If this were an actual application of the framework, Part B of Tool 3 would have to be filled out for each of the options selected in Step 2. For the sake of brevity, the sample shown here shows the answers to the Part B questions for just two options. Following that is an example of what the matrix might look like after working through Part B for all of the options checked in Step 2.

For the purpose of illustration, one option is selected from the matrix in Step 3 and is used to demonstrate Tool 4.A. A strategy package is chosen in Tool 4.B, the questions in Tool 5 are answered, and the matrix in Step 5 is demonstrated. No recommendation is made on whether to implement this strategy because that is outside the scope of this paper as an explanatory exercise.

Tool 1
Screening Criteria for Selecting Priorities

The responses below demonstrate how the questions in this tool could be answered based on the information that might be at hand before conducting an in-depth analysis. Possible sources would include the U.S. Environmental Protection Agency's (EPA) *Agenda for Action*, the U.S. Office of Technology Assessment's *Facing America's Trash*, and magazine articles.

1. Percentage Share of the Waste Stream

Americans buy over 3 billion household batteries a year. The University of Arizona's Garbage Project has estimated that the typical U.S. household disposes of 1.7 pounds of batteries a year. This amounts to over 83,500 tons a year for the country as a whole but is only a fraction of a percent of all municipal solid waste (MSW).

2. Expected Growth in Quantity and Share of MSW

Sales of household batteries have grown at 5 to 6 percent a year and are expected to continue to grow at this rate in the near future. The generation of MSW is projected to increase at 1 to 2 percent a year during the same period. Thus, batteries' share of MSW will be increasing by about 3 percent a year.

*Household batteries, paint, third-class mail, small appliances, and food-service disposables (single-service cups, plates, bowls, cutlery, etc.)

3. Toxicity

Household batteries contain various metals, including mercury, cadmium, lead, silver, mercury, nickel, zinc, manganese dioxide, and lithium. Mercury and cadmium are the primary concerns, because they are known to have adverse health impacts at low levels. Although substantial quantities of mercury and cadmium are used in household batteries in the aggregate, used batteries are widely dispersed throughout the waste stream. It is not clear whether the quantities or routes of exposure from household batteries present a threat. Attention has focused on the environmental impacts of incineration.

4. Wastes Generated Over the Product Life Cycle

No information identified at this time.

5. Special Handling Considerations

Some municipalities have begun using voluntary drop-offs to collect batteries for recycling. The benefits, economic feasibility, and safety of recycling have been questioned by some.

6. Availability of Information

Extensive information on the health impacts of mercury and cadmium is available. Access to information on the use of these materials in batteries and the alternatives is limited by its proprietary nature. There is very little research data on the threat posed by discarded batteries.

7. Identified Alternatives for Source Reduction

Several alternatives that use little or no mercury or cadmium are available. There are trade-offs in cost and performance, but ongoing research and development may improve the situation.

8. Other Criteria

No other criteria were selected for consideration.

Results of This Preliminary Screening

Given their use of toxic materials and projected rate of growth, as well as the availability of alternatives and the uncertainties regarding recycling, household batteries are good candidates for an in-depth analysis of source reduction potential.

Tool 2
Checklist of Source Reduction Options

See figure B.2 for application of Tool 2 to batteries.

Tool 3
Evaluation Questions for Selecting Options

The responses below demonstrate how the questions in this tool might be answered based on the information gathered during a more detailed literature search and on conversations with manufacturers, industry associations, and other knowledgeable individuals.

PART A. BASIC INFORMATION FOR ANY OPTION

1. Product

a. Models, Types, Sizes, or Styles

Household batteries can be characterized by at least three main distinctions: usage mode, shape and size, and chemistry.

Usage Mode: Single Use versus Rechargeable

"Primary" batteries are those which can be discharged only once, and are then thrown away. "Secondary," or storage, batteries can be recharged and reused hundreds of times. Most of the rechargeable batteries currently used in households are made of nickel and cadmium, and are known as "NiCads." The other type of rechargeable batteries currently sold for household uses are sealed lead-acid batteries.*

*Sealed lead-acid batteries will not be discussed here because the quantity used is smaller than NiCads, information is available, and the issues are similar to those faced by NiCads. Sealed lead-acid batteries have long been used in nonconsumer applications, such as emergency lighting, telecommunications, security systems, and power supplies for medical devices. However, they are finding increased use in consumer applications (such as portable tools, small appliances, televisions, lap-top computers, VCR cameras, etc.), and they are projected to continue to cut into the market for NiCads. Increasing attention will need to be focused on this category and its environmental impacts in the future.

Figure B.1
Tool 1 — Screening Criteria for Selecting Priorities
< Summary Matrix for Selecting Targets >

Criteria

Category	Contribution		Environmental Impacts		Other Considerations			
	% Share	Rate of Growth	Toxicity	Life-Cycle Impacts	Special Handling	Available Information	Feasible Alternatives	Other
Household batteries	*	**	***	?	? / ***	**	Yes	

Ranking System: *** = High, ** = Medium, * = Low, NA = Not Applicable, ? = Not known, x = See detailed description.

Figure B.2
Tool 2 — Checklist of Source Reduction Options

	Manufacturers	Consumers
1.	__ Eliminate product or reduce amount	X Don't purchase product or reduce use of product *Use fewer batteries; avoid battery-driven products where feasible.*
2.	X Eliminate or reduce toxic substances in the product *Reduce / Eliminate mercury and cadmium in batteries*	X Purchase product with reduced toxics *Buy batteries with less mercury and cadmium*
3.	__ Substitute environmentally preferred materials or processes	__ Purchase environmentally preferred products
4.	__ "Lightweight" or reduce volume	__ Purchase reduced products
5.	__ Produce concentrated product	__ Purchase concentrated products
6.	__ Produce in bulk or in larger sizes	__ Purchase in bulk or in larger sizes
7.	__ Combine functions of more than one product	__ Buy multiple-use products
8.	__ Produce fewer models or styles	__ Purchase fewer models or don't replace for style
9.	X Increase product life-span *If batteries can be made to last longer without a proportionate increase in toxicity, than fewer batteries will need to be purchased.*	X Purchase long-lived products *Purchase long-life batteries*

Producer	Consumer
10. __ Improve repairability	__ Maintain properly / repair instead of replace
11. __ Produce for consumer reuse	X Purchase reusable product / reuse product / donate to charity *Use rechargeable batteries*
12. X Produce more efficient product *For batteries, this is equivalent to #9 ("Long-lived Products") — A more efficient battery will last longer.*	X Purchase efficient product *Purchase efficient batteries*
13. N/A	X Use product more efficiently *Choosing the right battery for an appliance and caring for the batteries properly can reduce the number of batteries used.*
14. X Produce complimentary products *(A) Batteries will last longer in more energy-efficient battery-driven products.* *(B) Well-designed battery chargers can increase the life span of rechargeable batteries*	X Purchase preferred complementary products *(A) Purchase efficient battery-driven products.* *(B) Purchase well-designed battery chargers.*
15. __ Remanufacture products	__ Purchase remanufactured product
16. N/A	__ Borrow, share, or rent product
17. __ Other	__ Other

Figure B.3
Types, Sizes, and Applications for Household Batteries

Battery type	Principal sizes	Typical applications
Mercury-oxide	Button cells	Hearing aids, medical and photographic equipment, communication devices, pagers, and other applications requiring steady output voltage.
Silver-oxide	Button cells	Hearing aids, watches, photographic equipment, and specialty electronics requiring small, high-capacity batteries.
Zinc-air	Button cells	Hearing aids, medical monitoring instruments, pagers, communication devices, and other frequent use applications.
Alkaline-manganese	AAA, AA, C, D, 9-volt, and button cells	General purpose for flashlights, radio and television, tape recorders, toys, photographic equipment, watches, and instruments; popular for high drain applications.
Carbon-zinc	AAA, AA, C, D, and 9-volt	General purpose for flashlight, portable radios and electronics, toys, novelties, instruments, etc.
Lithium	AAA, AA, and button cells	General purpose in electronic and photographic applications requiring small, high-capacity batteries, such as watches, calculators, computers, memory backup, photo flash, motor-driven cameras and instruments.
Nickel-cadmium rechargeables	AAA, AA, C D and 9 volt	Portable tools, appliances, televisions, lap-top computers, memory backup, and portable electronics. (About 80% are sealed into appliances.)

Sources: Bell; Linden; National Electrical Manufacturers Association.

Shape and Size

"Cylindrical" batteries are used in flashlights, toys, radios, tape recorders, etc. The most common sizes (ranging from small to large) are AAA, AA, C and D. "Rectangular" batteries are available in lantern and nine-volt transistor sizes. "Button cells"—small batteries used in watches, calculators, hearing aids, etc.—come in a variety of sizes, voltages, and chemistries. Many other sizes and shapes also are used in medical, industrial, and other nonhousehold applications. Some batteries, particularly rechargeables built into appliances, have their size and shape custom-designed for a particular user.

Chemical Composition

The cylindrical and button cells on the market for household use are available in a variety of chemical compositions. The most common are carbon-zinc, alkaline-manganese (commonly known as alkalines), nickel-cadmium, mercury-oxide, silver-oxide, zinc-air, and lithium. Figure B.3 describes the principal sizes and typical applications for each of these battery types. There may also be subclassifications within a given category. For example, carbon-zincs are available in both general purpose and heavy-duty versions.

Batteries can also be classified by type of consumer and by the products in which they are used. In 1988 battery users and their purchases were: federal government, $120 million; original equipment manufacturers (OEMs), $162 million; industrial users, $260 million; retail button cells, $420 million; and retail cylindrical cells, $2.626 billion (figure B.4).[1] The proportion of environmental damage is not necessarily related to sales value, since different classes of consumers use different battery types and disposal methods. The mercury used in 1988 in mercury-oxide batteries, which are just one of the types of batteries which use mercury, was distributed among the following applications: 10 percent, industrial; 14 percent, medical; 27 percent, consumer; and 48 percent, military.

Figure B.4
1988 U.S. Battery Market ($ U.S. Millions)

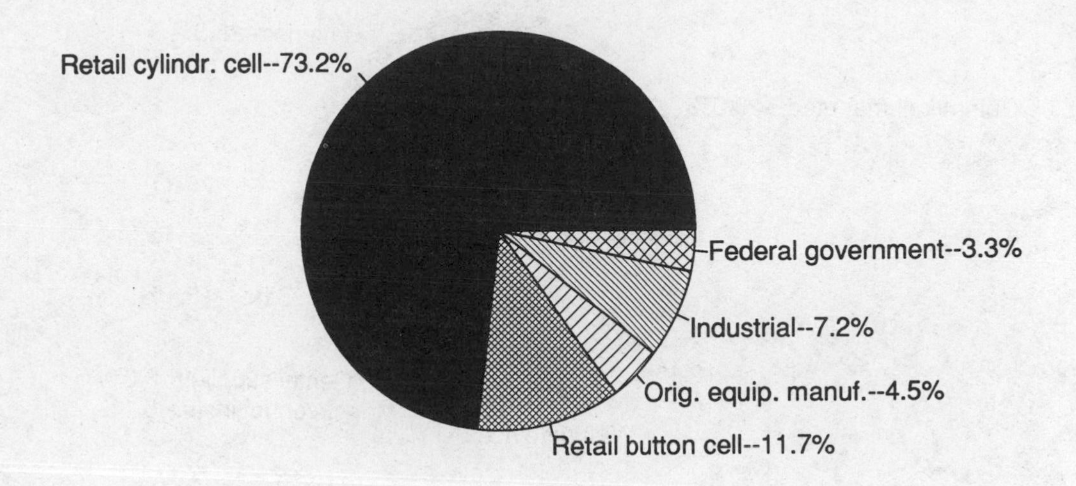

Source: *Chemical Business.*

Figure B.5
Consumer Battery Sales, by Uses, 1987

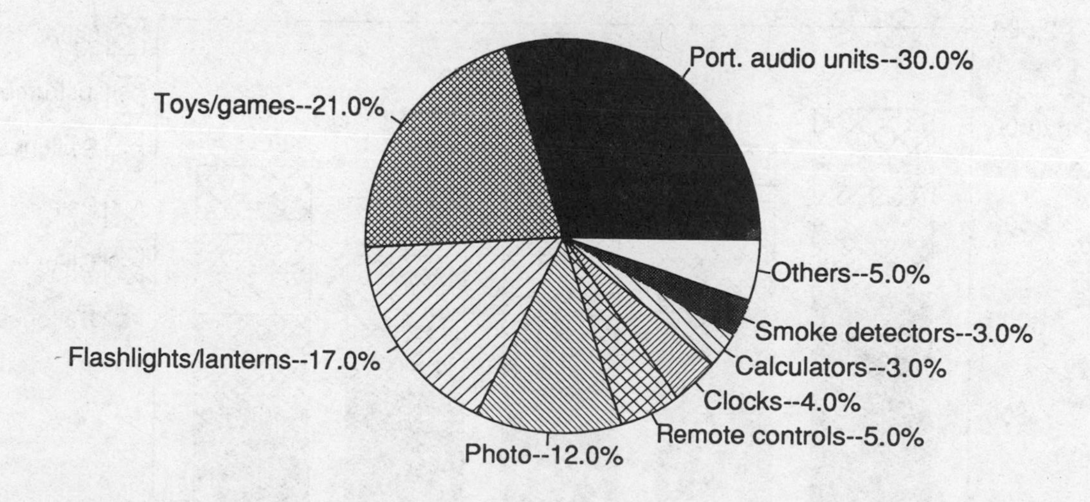

Source: *Drug Store News.*

Figure B.6
Mercury Consumed in the United States, 1988

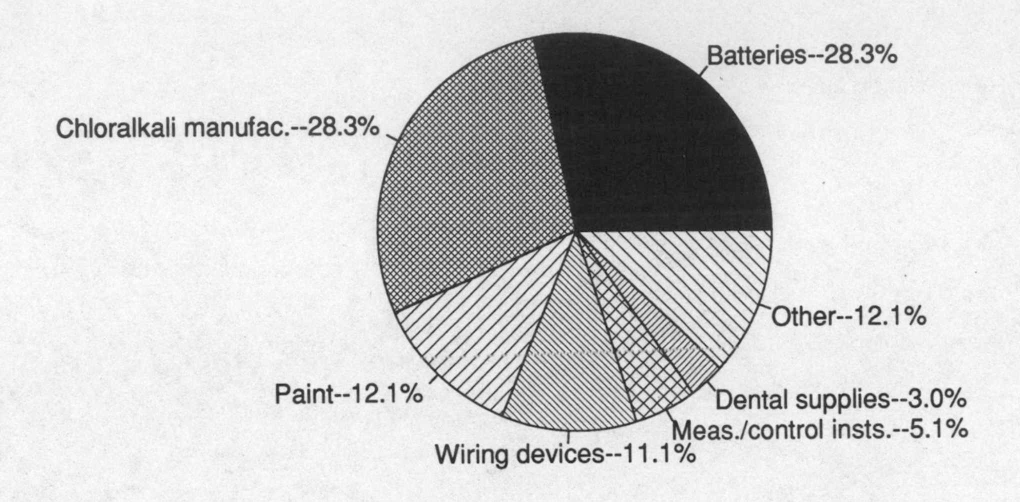

Source: Bureau of Mines.

Figure B.7
Mercury Consumed in the United States

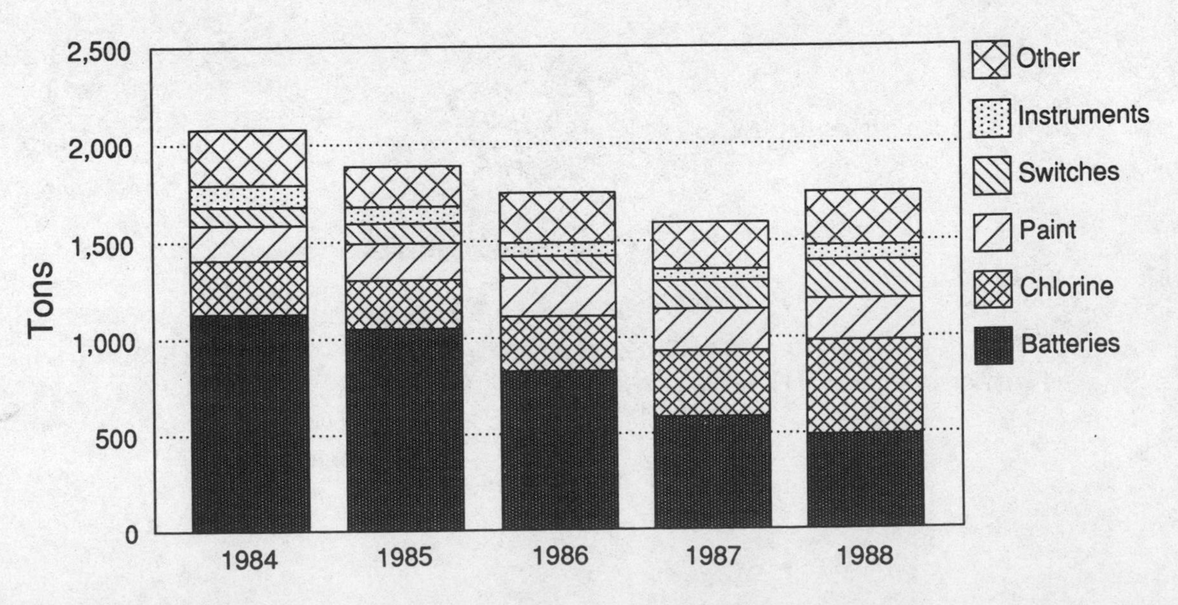

Source: Bureau of Mines.

The average household used 32 batteries a year in 1988. But some heavy-use households buy up to 92 batteries a year. These households are typically young married couples, with children ages 6 to 12 or 13 to 20.[2] End-uses for household batteries were estimated in 1987 as follows: portable audio units, 30 percent; toys and games, 21 percent; flashlights and lanterns, 17 percent; photo, 12 percent; remote controls, 5 percent; clocks, 4 percent; calculators, 3 percent; smoke detectors, 3 percent; and other, 5 percent (figure B.5).[3]

b. Quantity of the Product

U.S. battery sales for 1990 were expected to be over 3 billion. Sales of alkalines were projected at 1,928,000,000 and carbon-zincs at 710,000,000.[4] Household sales of nickel cadmium rechargeables were estimated at 174,000,000.[5] Sales figures for button cells were not identified.

Amounts of Metals

Major uses of mercury include batteries, chlor-alkali manufacturing (for the production of chlorine gas and caustic soda), as a preservative in latex paint,* in electrical wiring devices and switches, and in measuring and control instruments. According to the Bureau of Mines, the battery industry used 493 tons of mercury in 1988.[6] This was 28 percent of mercury consumed in the United States for all uses (figure B.6). This was down from 1,129 tons, and 54 percent of mercury consumption, in 1984 (figure B.7).† However, not all of this mercury was used in household batteries. According to the National Electrical Manufacturers Association (NEMA),‡ the amount used in *consumer* batteries fell from 778 tons in 1984 to 225 tons in 1988, and the level has continued to drop (figure B.8) Note that these figures do not account for imports or exports.

*Use of mercury as an additive to latex paint will be reduced based on a voluntary agreement signed between EPA and manufacturers in June 1990.

†These figures do not account for another major source of mercury in the environment, the combustion of coal for electric power.

‡The industry associations operate out of the National Electrical Manufacturers Assoociation. The NEMA Dry Battery Section represents domestic manufacturers of dry cells, both primary and rechargeable. The Battery Products Alliance (BPA), formed in June 1989, addresses environmental issues of rechargeable batteries. It is composed of foreign and domestic manufacturers and distributors of rechargeable batteries and products that use those batteries. In addition to battery manufacturers, BPA members include Black and Decker, General Electric, Motorola, Panasonic, the Power Tool Institute, Sanyo, and Skil Corporation.

Cadmium is used in NiCad batteries, for metal plating for corrosion resistance, in pigments, and as a heat stabilizer in plastics. It is also introduced into the environment by fossil fuel combustion and fertilizer use. Although the Bureau of Mines does not collect data on cadmium usage, it estimated that the battery industry as a whole used 32 percent of the cadmium consumed in the U.S. in 1988.[7] Again, this does not include emissions from power plants or fertilizer use.

Approximately three-quarters of the NiCads sold are for household use.[8] A report prepared for EPA by Franklin Associates estimated that battery discards into MSW in 1986 included 930 tons of cadmium, accounting for 52 percent of cadmium in MSW (figure B.9).** Franklin projected discards of cadmium in batteries to increase to 2,035 tons (and 76 percent of cadmium in MSW) by the year 2000 (figure B.10).

Not all of the mercury and cadmium used in batteries is used in household batteries. Much of it is used in specialized military, industrial, and scientific applications, which are outside the scope of this paper. However, these other types of batteries can still have environmental impacts.

Mercury and cadmium are not the only metals used in batteries. One study estimated that 4,400 tons of nickel, 18,500 tons of zinc, and 37,400 tons of manganese dioxide were used in household batteries in 1987.[9]

c. Function of the Product or Material of Concern

Mercury and cadmium are the materials of primary concern in batteries. Figure B.11 presents estimates of the levels and amounts of mercury and cadmium in household batteries. (Note that these estimates are from different sources and are not necessarily comparable.)

Battery manufacturers and product designers face a complex set of trade-offs between such factors as size constraints, shelf life (how long the battery lasts when not in use), voltage, current, capacity, power density (capacity-to-weight and -size ratios), frequency of use, drain rate, cost, response to temperature changes, shape of the discharge curve, change in discharge with use, and cycle-life (for rechargeables). Choices of which materials to use in a battery, and what battery to use in a product, are made on the basis of these trade-offs.

Mercury-oxide button cells have relatively steady

**This estimate of *disposal* is higher than the Bureau of Mines estimate of *usage* because it includes net imports and because batteries have a shorter life span than some other products that use cadmium.

discharge characteristics, a high capacity-to-volume ratio, good high-temperature characteristics, and good resistance to shock, vibration, or acceleration.[10] These qualities make them suited for use in numerous products such as hearing aids, calculators, and watches. The mercury-oxide acts as the positive electrode, and thus the level cannot be reduced without reducing the capacity of the battery. Mercury-oxide batteries used 116.4 tons of mercury in 1988, 46.8 tons of which were in button cells used by consumers.[11] But, since mercury-oxides are often recycled, they probably contribute a relatively lower share of the mercury from batteries entering the environment.

The U.S. market for mercury-oxides totals $195 million, of which $100 million is in hearing aids and $60 million in watches. Mercury-oxide batteries currently have about 40 percent of the button-cell market for hearing aids. The number used in hearing aids is growing at about 4 percent a year, but the share is rapidly declining as they are being replaced by zinc-air batteries.[12] There will probably be a shift away from mercury-oxides to lithium and zinc-air batteries in other uses as well.

The other types of batteries that contain mercury have much lower levels, of a few percent or a fraction of a percent. Mercury is used to coat the negative electrode and acts to prevent the formation of bubbles of hydrogen gas. Such gassing lowers the voltage available from the battery and shorten its shelf life. Gas buildup may also cause leakage and corrosion by breaking the battery's safety vents (which are designed to prevent an explosion due to excessive gas pressure). Attention has centered on the use of mercury in alkaline batteries because of their high sales volume.

Nickel-cadmium batteries are the main source of cadmium from batteries. (Cadmium also occurs as an impurity in the zinc used in primary batteries.) NiCads are 18 to 22 percent cadmium. NiCads' advantages include long cycle-life, good low-temperature and high-rate performance capability, and long shelf-life in any state of charge.[13] Because cadmium is used as the negative electrode in NiCads, the amount cannot be reduced without reducing battery capacity.

The National Electrical Manufacturers Association (NEMA) estimates that 75 to 80 percent of NiCad batteries used by consumers are sealed into appliances. The remaining 20 percent of NiCad batteries, purchased by consumers in blister packs in drug and hardware stores, represents only 1 percent of household batteries that are purchased and disposed of. This share is likely to increase in the future. More appliances will be built with rechargeable batteries that are removable, in order to facilitate recycling.

d. Toxicity

Household batteries are an issue in MSW disposal because of the toxicity of the metals they contain. Household batteries are made of a variety of materials, including zinc, manganese dioxide, mercury, cadmium, silver, and lithium. Concern has focused on the use of mercury and cadmium, because they are toxic at low levels and because batteries are major sources of these elements in the waste stream.

Mercury can damage the kidneys and the central nervous system (producing fatigue, rash, cramps, fever, personality changes, memory loss, and delirium) and can cause fetal damage. Human exposure to mercury is most likely to occur through fish consumption, since the bioaccumulation factor in fish is from 10,000 to 100,000 the level in ambient water. According to one estimate, mercury levels of two parts per trillion in lake water may result in fish consumption advisories.[14] Thus, people who consume large quantities of fish are at increased risk. Fetuses are the most sensitive population, even though their mothers may show no adverse effects.

Cadmium is a probable human carcinogen and can cause lung and kidney disease. The ingestion of food is the prime pathway, since cadmium is bioconcentrated in grains and cereal products, and animals that feed on them, as a result of fertilizer use and disposal of sewage sludge. Cadmium residues in the general population are reaching maximum safe levels, and thus all exposure pathways are of importance when controlling for exposure. Populations at risk include smokers, people living near smelters and incinerators, those who eat organ meats, fish, grains, and vegetables from cadmium-enriched soil and water, and those using acidic drinking water. Various factors can affect sensitivity. Besides the human health risk, both mercury and cadmium may have other environmental effects, such as on aquatic organisms.

Use of other metals (including zinc, manganese, and nickel) raises potential human health and environmental concerns. However, knowledge of the health effects of these metals is based largely on occupational exposure at high levels. Data on the effects of the low exposure levels (such as would be attributed to the use of these metals in batteries) are very sparse, so these metals will not be discussed further.

The effects of battery disposal are not well known at this time. Exposure could come from incinerator

Figure B.8
Mercury Used in U.S. Consumer Batteries.

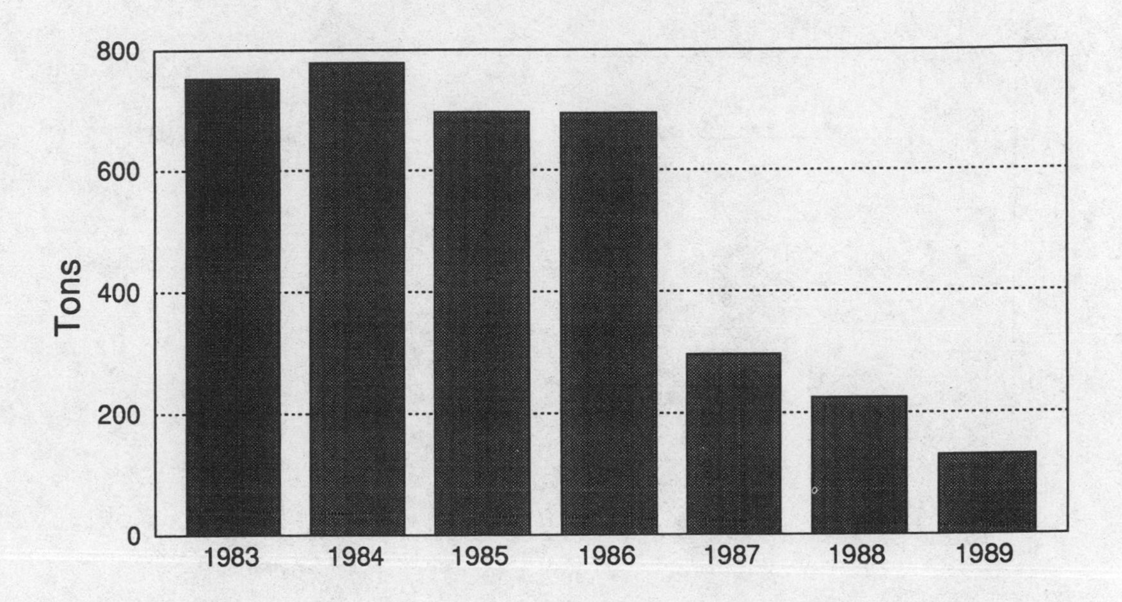

Source: National Electrical Manuacturers Association.

Figure B.9
Cadmium Discarded in Municipal Solid Waste, 1986

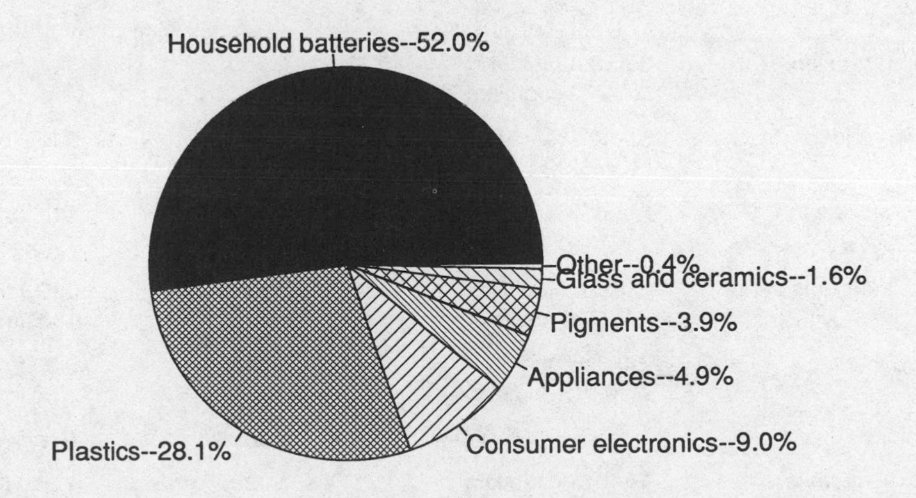

Source: Franklin Associates.

Figure B.10
Cadmium Discarded in Municipal Solid Waste

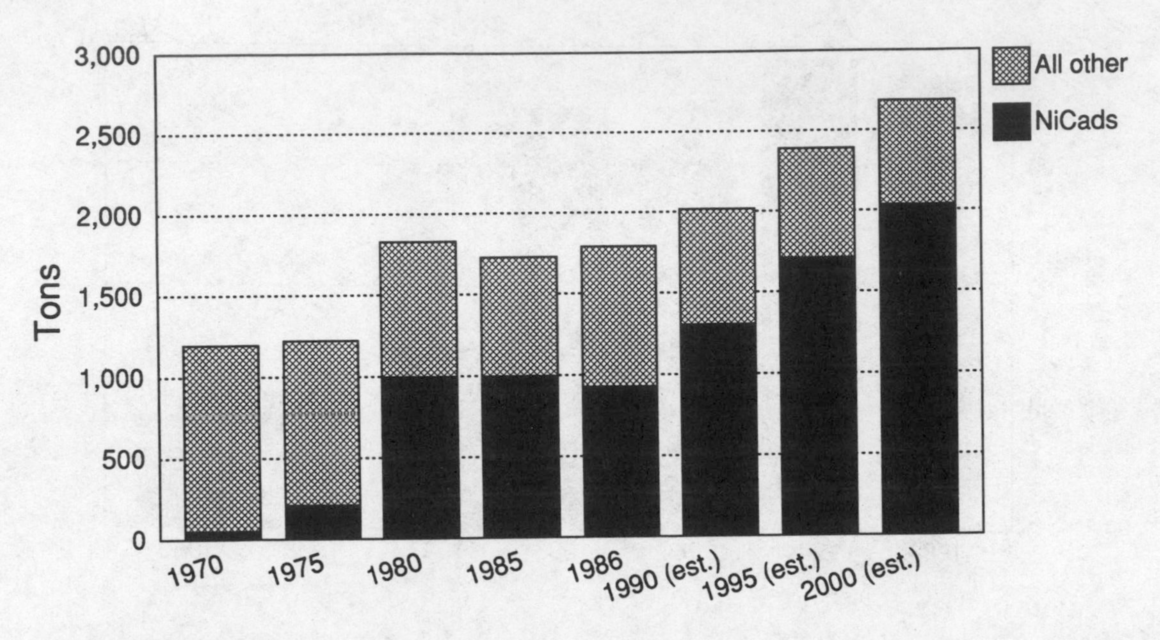

Source: Franklin Associates.

Figure B.11
Mercury and Cadmium in Household Batteries

Battery type	Typical Mercury or Cadmium Weight per Cell (%)	Consumer Sales	Estimated Quantity of Mercury or Cadmium (Tons)
Mercury-oxide	35-50% mercury .3-.45% cadmium	Not available	32.6 tons mercury
Silver-oxide	About 1% mercury .3-.45% cadmium	Not available	0.003 tons mercury
Zinc-air	About 2% mercury 1.1-1.22% cadmium	Not available	1.05 tons mercury
Alkaline-manganese	< 1% mercury .25-.54% cadmium	1,928,000,000	256.8 tons mercury 16.5 tons cadmium
Carbon-zinc	< 0.01% mercury .366-.6% mercury	710,000,000	3.52 tons mercury 20.6 tons cadmium
Lithium	None	Not available	None
Nickel-cadmium rechargeables	18-22% cadmium	174,000,000	1,200 tons cadmium *(in appliances)* 262.5 tons cadmium *(sold separately)*

Note: These estimates are from different sources and may not be comparable.

Source: Carnegie-Mellon; Minnesota Pollution Control Agency; National Electrical Manufacturers Association.

emissions, incinerator ash, or landfill leachate. At this point, the risks from any of these sources are uncertain, but there is some information available.

Mercury is difficult to control in incinerator emissions because of its volatility. Pollution control equipment can remove 45 to 85 percent of the mercury in the flue gas, depending on the technology.[15] As would be expected, given that batteries have historically been the largest consumers of mercury, several foreign and domestic studies have shown that removing household batteries from the waste stream reduces mercury emissions from incinerators by 60 to 80 percent.[16] However, the studies do not appear to have addressed the actual risk from these emission levels. Studies have estimated that direct inhalation of mercury emissions from incinerators creates low risks to humans. However, precipitation will deposit atmospheric mercury into lakes, where it will bioaccumulate in fish and may create a risk to humans.

Incineration also produces incinerator ash. If mercury batteries are incinerated, some of the mercury will be captured in the incinerator ash. The rate of capture depends on the type of pollution control device being used. Because cadmium is not as volatile, nearly all of the cadmium is likely to be captured in the ash.[17] Tests of incinerator ash have found that samples often exceed EPA's extraction procedure (EP) toxicity test for cadmium. Depending on the final regulatory status of incinerator ash this could limit beneficial uses of the ash or force it to be disposed of as a hazardous waste. Actual leachate collected from ash monofills may fail drinking water standards but generally does not exceed the EP toxicity levels.

Industry studies done in the past have shown that discharged alkaline and carbon zincs sometimes passed the EP toxicity test and sometimes failed it.[18] An independent study of NiCads found that they exceeded the levels in the EP toxicity test when their casings were damaged.[19] Two studies have shown that disposing of batteries in landfills does not appear to present a threat in the short term, because the metals are not likely to escape from the batteries very quickly.[20] The long-term fate is less certain. Whether mercury and cadmium eventually enter groundwater would depend on whether landfills are designed adequately to contain the leachate and on the degree to which the metals bind to the soil (which in turn depends on the type of soil and the properties of the leachate).

e. Intended Life of Product and Trends in Average Life Span

A battery's life span depends on the type of product that it is being used in. For example, a battery that lasts 100 hours when used in a low-drain application for 4 hours a day may only last 20 hours when used continuously in a high-drain application.[21] In general, battery life span has improved in recent years.

Batteries may be disposed of before the end of their useful life. Franklin Associates estimated that consumers discard half of all NiCads sold in blister packs within the first three months of purchase, because they become frustrated with recharging them.[22] (The remaining batteries are estimated to last up to four years.) NiCads sealed into consumer products will also be discarded when the appliance is thrown away, whether or not they are still working.

Manufacturers have continually been able to improve the performance of the various battery types. Eventually, though, the limits will be reached, and producers will have to switch to different chemistries to improve performance or reduce toxicity.

f. Economic Life Cycle and Trends

Some battery types are at the innovation stage (lithium); others are mature (alkalines); still others are being replaced by alternatives (carbon-zinc, mercury-oxide).

g. Major Alternatives

Mercury Alternatives

For *alkalines* and *carbon-zincs*, the alternative is to substitute a different material than mercury to control gassing. Manufacturers are continuing work toward reducing mercury levels. They are using the experience they have gained in Europe to introduce low-mercury batteries in the United States, with the goal of producing mercury-free batteries. The industry is also searching for substitutes that perform the same function as mercury, including indium, thallium, and organic compounds.[23] Information on the full range of substitutes is proprietary, so it is difficult to estimate the overall effect on waste-stream toxicity. In addition, there is less data on the environmental effects of the substitutes than on the material they are replacing.

The options for *mercury-oxides* involve switching to different battery chemistries that require little or no mercury. Alternatives include silver-oxide, zinc-air, and lithium.

Silver-oxide batteries can substitute for mercury-oxides in a number of applications. Their performance advantages include higher voltage, greater capacity, and superior low-temperature capacity. However, they can be considerably more expensive than mercury-oxides.[24]

Zinc-air button cells are replacing mercury-oxides in many applications. Zinc-air cells use oxygen taken from the air as the active cathode material. The air intakes are covered with a seal, and the cell is activated by removing the seal. Once started, this reaction cannot be stopped. This makes zinc-air batteries unsuitable for products used only intermittently. They work best in frequently used products with moderate drain rates, such as hearing aids. They are not well suited for low-drain applications such as watches, because they would deteriorate before using their full capacity.[25] They can sell for up to twice as much as equivalent mercury-oxide button cells, but they can last twice as long in some applications.

Lithium batteries can substitute for mercuric-oxide button cells and small alkalines. There are a number of lithium chemistries available or under development. Lithium batteries have high energy density and a long shelf-life (up to 10 years). They have been produced in various sizes, most commonly in button cells, AA, and AAA. One use has been in computers to provide current to memory chips if the power fails. They are also well suited for high-drain applications such as automated 35-mm cameras. Manufacturers are working on developing lithium batteries in popular consumer sizes, although safety issues may limit their size.

Most lithium metals are highly reactive with air and water. Lithium batteries have safety features to prevent them from catching fire or exploding if short circuited, and their safety record has been good.[26] Questions have been raised, however, about whether they may present problems in collection and recycling programs. (Some programs deactivate them before disposal.[27]) Manufacturers are currently working to develop lithium batteries without lithium metal, which should be inherently safer. Lithium batteries are projected to increase in popularity, and their environmental implications deserve further study.

Cadmium Alternatives

The options for reducing cadmium use also involve switching to different battery types. One alternative is to increase the usage of sealed *lead-acid* batteries, which are already used in a limited number of household applications. This is not source reduction, however, since lead is also toxic. The advantage of lead is in the ease of recycling. There are no domestic recyclers of NiCads at this time, although some material is sent overseas for reclamation. By contrast, an infrastructure for lead recycling does exist (since most lead-acid car batteries are recycled). However, there are questions about the environmental records of secondary lead smelters. And lead-acid batteries cannot be recharged as many times as NiCads, so they are not as effective at conserving resources.

Rechargeable lithium batteries were seen at one time as having the potential to replace NiCads in some applications. Unlike lithium primary batteries, lithium rechargeables had serious safety problems, and production was halted. New, safer types of lithium rechargeables are under development, but safety will continue to be an even bigger issue in lithium rechargeables than in lithium primary batteries. Even when the safety problems are solved, rechargeable lithiums will probably be limited to expensive products such as lap-top computers, because they require special circuitry to control their charging and discharging.

The battery industry sees *nickel metal hydride* rechargeables, currently being developed, as the best option to reduce cadmium use. These batteries will replace cadmium with various rare earth metals. One metal hydride battery under development uses a vanadium-zirconium-titanium-nickel-chromium alloy as the negative electrode, nickel hydroxide as the positive electrode, and potassium hydroxide as the electrolyte.[28] Other possible elements include lanthanum, cerium, cobalt, and neodymium.[29]

Nickel hydrides may be introduced for applications requiring low to medium drain rates (such as lap-top computers and VCR cameras) by 1991. Nickel hydrides have not yet attained the high drain rates required for use in appliances such as power tools. It is likely to be several years before nickel hydrides are commercially available for high drain applications.

Another set of alternatives for both mercury and cadmium is to use nonbattery alternatives such as solar cells and super-capacitors. However, those alternatives will not be addressed here.

h. Market Concentration

The household battery market is dominated by Eveready and Duracell, with intense competition between them. The industry leaders are followed far behind by Rayovac and Kodak (which is trying to break into the market). House brands and rechargeables account for a small share of the market. Industry market shares in November and December 1989 for alkaline batteries sold in food, drug, and mass merchandise outlets were 41.5 percent for Duracell, 35.9 percent for Eveready, 8.7 percent for Rayovac, and 7.6 percent for Kodak.[30]

The major marketers of rechargeables are Gates Energy Products, SAFT America, and Panasonic. Rechargeable batteries produced by other companies are imported built into appliances. In addition, numerous other companies are specialized and sell to military, industrial, or medical users.

Domestic battery manufacturers have been owned by corporations that make other products. Ralston Purina Co. bought Eveready from Union Carbide Corp. in 1986, and Duracell was owned by Kraft prior to a leveraged buyout by Kohlberg, Kravis, Roberts & Co. in 1988.[31] But, unlike foreign companies, the parent companies in the United States do not make battery-driven products. In Japan, Matsushita (Panasonic), Sanyo, Hitachi, and Casio produce both batteries and products that use them.[32] This allows them to closely coordinate design decisions by both battery and appliance designers.

i. Trends in Product Design, Use, Distribution, and Pricing, and Factors Influencing Those Trends

Less Mercury

Mercury levels in alkalines have dropped from approximately 5 percent to under 1 percent, and as low as 0.025 percent, in the last decade.

Longer Life / Lower Costs

From 1982 to 1987, battery manufacturers were able to extend the service life of alkalines by over 40 percent and reduce the price.[33]

Shifts among Battery Types

Alkaline batteries have increased their market share, replacing carbon-zincs. Alkalines, which are popular with consumers because of their long life, are expected to account for 73 percent of the primary cylinder battery market in 1990.[34] This shift from carbon-zincs to alkalines is expected to continue.

In other battery types, mercury-oxide button cells are being replaced by zinc-air and lithium batteries. Once nickel hydrides and rechargeable lithiums are introduced they will begin replacing NiCads.

Smaller Sizes

There has been a shift toward smaller batteries in response to the increasing use of sophisticated electronics. AAs account for nearly 60 percent of the alkaline market, and AA and AAA are projected to continue to experience the fastest growth in the market. Eveready is even introducing a AAAA size. There are two reasons for the shift to smaller sizes. The new electronics mean that the efficiency of many products using batteries has increased, so less capacity is needed. And as products are miniaturized, the physical space available for batteries has decreased. The increase in efficiency also means that shelf life has become as important as how long it lasts in use.

More Uses

An increasing number of products have been designed to run on battery power. Examples include portable vacuum cleaners, power tools, digital tape measurers, and portable televisions (to name just a few).

Change in Casing Materials

Another change has been the switch from steel to plastic casings for primary batteries. Because the plastic is thinner, more of the "active materials" can be put in a given sized battery, increasing the life of the battery. This is particularly important in the AA and AAA sizes, because the case makes up a proportionately larger share of the item. This is a form of source reduction, since less material must be used in the shell than in the past.

Changes in Appliances

Changes in product use can also affect battery development and sales. Two major trends in appliances have been miniaturization (requiring smaller batteries) and the addition of more features (requiring more power). The popularity of portable cassette players contributed to the growth in sales of AA batteries. And the evolution of lap-top computers forced manufacturers to develop smaller, lighter batteries to suit the needs of the computers.

Design Decisions

At one time, design decisions for batteries were made separately from design decisions for the products that used them. Designers for original equipment manufacturers (OEM) either developed products based on off-the-shelf batteries or conceived new products and required battery manufacturers to fabricate batteries that met their specifications. Either way, there was not enough interaction between the two industries. This is changing somewhat. Joint ventures and nondisclosure

Figure B.12
European Efforts at Source Reduction in Household Batteries

Austria
Planned to implement a labeling and deposit collection system to segregate batteries from MSW.

Denmark
Planned on implementing a mandatory rebate of approximately 10 percent on NiCad batteries to encourage their collection. The rebate would include NiCads sold individually and sealed in appliances.

Germany
Industry has agreed to reduce mercury levels to 0.1 percent and to label mercury-oxides, alkalines, and NiCads that exceed this level.

Netherlands
Collects all types of batteries, although only silver-oxide and NiCads are recycled. Battery manufacturers and importers have agreed to reduce the mercury in alkaline batteries to 0.15 percent of current levels.

Norway
Planned to implement a tax of 30¢ to 45¢s on batteries containing more than 0.025 percent mercury and cadmium.

Sweden
Prohibits sales of batteries containing more than 0.025 percent mercury and/or cadmium and disposal into MSW of such batteries purchased before ban went into effect in 1990.

Switzerland
Requires household batteries containing more than 0.025 percent mercury and/or cadmium to be labeled as hazardous substances. Considering a deposit system to increase recycling rates.

Commission of European Community
The European Community will limit alkaline batteries to be used under normal conditions to 0.025 percent mercury by 1993 and those used under extreme conditions to 0.05 percent mercury. Button cells will be exempt from this requirement. Countries will also set up recycling programs.

European Dry Cell Manufacturers Association
Has committed to reduce mercury levels in cylindricals to 0.025 percent by 1992 and to substitute zinc-air cells for mercury-oxides wherever possible.

Sources: Forker, Bureau of National Affairs, Hershkowitz

agreements have become more common, particularly for products like lap-top computers (where development time for new products has been decreasing, and designers are continually demanding smaller batteries with a longer life).[35] While these interactions are more likely to occur than in the past, more can still be done. According to one battery manufacturer, "Too many times we see a new product for which we might have been able to design a more cost-effective or more efficient battery, if a designer had simply phoned us. Joint efforts pay off."[36]

2. Source Reduction in the Industry / Firm

a. Source Reduction Experience by Manufacturers

Whereas alkaline batteries contained up to 5 percent mercury until a few years ago, the mercury content in U.S. alkaline batteries is now well under 1 percent, and is approaching 0.025 percent mercury in some batteries. Thus, despite a continual increase in the number of batteries sold, the total quantity of mercury used has declined. As shown in figure B.8, the amount of mercury used in domestically produced *household batteries* fell from 694 tons per year in 1986 to 131 tons per year in 1989, and the trend is projected to continue. (These figures do not include medical,

Figure B.13
Eveready's Mercury Reduction in Household Batteries, late 1980s*

Battery size	U.S. Eveready	European Eveready
D cell	83	98
C cell	78	98
AA	80	96
9 volt	90	93

*Eveready has since made further mercury reductions in the United States.

Source: Minnesota Pollution Control Agency.

military, or industrial batteries or imported household batteries.) At the same time, manufacturers were able to extend service life and reduce the price.

Manufacturers were able to achieve this reduction both by eliminating the need for mercury and by switching to substitutes for mercury. In the first instance, they reduced the impurities in the other materials in the batteries, redesigned the venting systems so that there was not as much of a problem with gassing, and found that better production techniques in applying the mercury could drastically reduce the quantity used.[37] (Fine spreading reduced the quantity of mercury required by up to 95 percent.[38]) In the second case, the industry began searching for substitutes that perform the same function as mercury, such as indium, thallium, and various organic compounds.

Manufacturers in Europe, including some of the same companies operating in the United States, have been able to achieve greater reductions in mercury abroad. This has been possible because European consumers have been willing to accept tradeoffs in battery performance in return for decreased mercury levels.[39] Producers have built their sales in the United States on long life and reliability and may be unwilling to accept any constraints on performance to reduce mercury levels. Thus, it has taken them longer to reduce mercury levels in the United States.

b. Factors Stimulating Past or Present Source Reduction

Mercury levels have received more attention in Europe than in the United States. In 1986 Switzerland required household batteries containing more than 0.025 percent mercury and cadmium by weight to be labeled as hazardous substances. The Swiss law has been credited with spurring reductions in mercury and cadmium levels. In the past few years, several European countries have followed suit. Figure B.12 lists the actions various European countries are taking.[40] Figure B.13 shows the mercury reduction in European and U.S. batteries achieved by 1987 and 1988.[41]

Regulations on mercury discharge in wastewater and occupational exposure have provided an incentive to reduce mercury levels in the United States. Doing so has also saved manufacturers money, since they buy less mercury. (Mercury substitutes may be more expensive, so to the extent that manufacturers must switch to substitutes, costs may increase.) If adopted, a proposal by the U.S. Occupational Safety and Health Administration in February 1990 to reduce permissible exposure limits for cadmium by up to 99 percent may provide an additional incentive to develop substitutes for NiCad batteries.

A Minnesota bill on batteries, H.F. 1921, was signed into law in April 1990. One of the provisions requires that by January 1, 1992, no manufacturer will be able to sell or distribute alkaline batteries containing more than 0.025 percent mercury (the same level as required by several European countries). The law may further speed source reduction in the United States.

c. Implications for Future Efforts

Mercury levels in alkalines and carbon-zincs have dropped dramatically in recent years. This would seem to indicate that manufacturers have already exploited the easy reductions and that elimination of mercury will be more difficult. But battery manufacturers feel that they can eventually eliminate mercury from these batteries.

Previous source reduction efforts in button cells and rechargeables have been aimed at developing alternative chemistries. These alternatives have not yet been perfected, but previous efforts put development that much closer.

Source reduction in batteries can be both deliberate (reducing mercury and cadmium usage) and inadvertent (increasing battery life, or improving the efficiency of battery-driven products). Interestingly, battery manufacturers have traditionally viewed source reduction as involving only toxicity reduction (reducing the level of mercury, or developing rechargeable batteries that do not contain cadmium). They have not viewed increasing battery life or improving the efficiency of battery-driven products as being source reduction, even though this results in fewer batteries being used in a given application, and thus fewer batteries being discarded in MSW. Of course, if batteries last longer in a given application, consumers might increase their use of battery-driven products enough that more batteries would be discarded into landfills.

3. Consumer Behavior

a. Consumer Preference

Performance (long life) and price are the key aspects to consumers.[42] However, the key characteristics will vary depending on the type of battery and product. For instance, a 1984 study for EPA found that "the requirements for a flashlight battery are: low cost, long shelf-life, suitability for intermittent use, and moderate operating life."[43] The third characteristic is less important in the case of a hearing aid or watch battery, and a *long* operating life is more important.

Price seems to be less important than performance. According to a store manager, "Batteries move very well, and price doesn't matter that much. People are going to buy them whether we get them on deal or not, because they need them."[44] A 1984 study for EPA found that the demand for batteries, as an industry, is generally inelastic—that is, price levels have very little effect on demand for batteries. This was attributed to the lack of substitutes for batteries as a group. Somewhat higher elasticities were found for individual battery product groups because consumers are often able to substitute one battery type for another. (This effect will vary by use. Sales volume will be much more closely linked to price in those uses where corded appliances can be substituted.) It is not clear what effect a significant price increase (whether due to product reformulation, change in chemistry, or a tax) would have on demand.

Marketing is seen as critical in determining consumer's preferences. According to an industry executive, "Shoppers won't go across the aisle for a competitive battery. From their viewpoint, batteries are just above shoelaces in interest. They don't twinge at the heartstrings, so we try to make up for that with advertising."[45] Marketing has focused on length of life.

b. Consumer Misuse

Consumers can inadvertently shorten battery life through improper use and storage. Recommendations by battery manufacturers to prolong life include the following[46]:

- Batteries, especially carbon zincs, should not be stored in hot places.
- Old and new batteries should not be used in the same appliance, because this will draw down the new battery faster. All the batteries in an appliance should be changed at the same time.
- Battery types, such as alkalines and carbon-zincs, should not be mixed in the same appliance.
- Cleaning the battery terminals with an emery board, pencil eraser, or coarse cloth may keep them functioning longer.

c. Source Reduction Experience by Consumers

Until very recently, none of the major U.S. battery manufacturers labeled or advertised their products on the basis of mercury levels. Thus, there is little evidence about consumer preferences toward low-mercury batteries. Manufacturers feel that, unlike Europeans, consumers in this country are not willing to change their demand for the highest performing battery at the lowest possible price. If this is true, it will limit the opportunities for source reduction to those that do not impede performance or increase price.

Despite their availability, NiCads account for a small share of the household battery market. Many consumers feel that the savings that can be achieved by using NiCads are not worth the inconvenience of their short cycle-life and the need to recharge them. As noted earlier, it has been estimated that half of the NiCad batteries sold in blister packs are discarded within three months of purchase. Still, the number of NiCads sold has increased in recent years and is expected to continue to grow.

4. Waste Management

a. Disposal Patterns

Consumers generally throw out batteries with ordinary trash. Given that 73 percent of MSW in the United States is landfilled, it is likely that the majority of batteries enter MSW landfills. Battery packages carry a warning not to burn them, but consumers are generally not aware of whether their garbage is landfilled or incinerated. Even if they know, they may not have an alternative disposal option.

Some batteries are collected in household hazardous waste collection programs, where they may be recycled, shipped overseas for recycling, or disposed of in hazardous waste landfills. In addition, many jewelry stores, watch repair shops and other retail stores that replace button cells for their customers collect the batteries and sell them to recyclers.

In 1984 EPA estimated that 1 to 5 percent of batteries are rejected by manufacturers at the plant. Other estimates put the figure closer to 1 percent. Still, given that there are roughly three billion batteries produced a year, this could lead to large quantities of batteries being disposed of in bulk. Waste handling practices differ by manufacturer and battery types. Depending on the manufacturer, alkaline batteries may be sent to hazardous waste landfills or local municipal landfills, or disposed of on-site.[47]

b. Recycling Considerations

As the following discussion suggests, while source reduction is the best long-term solution for reducing environmental impacts from household batteries, recycling may be the best short-term method for NiCads and certain button cells.

Currently, only mercury-oxide and silver-oxide buttons cells are economical to recycle, and only the mercury and silver themselves are recovered—the rest of the battery is disposed of. Although battery recycling programs around the country collect batteries of all compositions, the other types of batteries are sent to hazardous waste landfills. The mercury content in carbon-zinc and alkaline batteries is too low to profitably recover using current processes, and it will become less economical as the mercury levels continue to decline. The cost of purifying the other materials in carbon-zinc and alkaline batteries is currently higher than the cost of virgin materials.[48] There are processes available that could reclaim a wider variety of materials from a wider variety of batteries, but they are commercially unproved.[49]

NiCads have a high enough percentage of cadmium, which has a sufficiently high market value, that recycling might be feasible (depending on the price of virgin material[50]). But no domestic facilities for cadmium recycling are currently in operation. (There is a facility that recovers nickel, which constitutes 15 to 25 percent of the battery, for making stainless steel.[51]) Plant scrap, industrial batteries, and possibly some of the NiCads collected in recycling programs are sent to overseas facilities for recovery of the cadmium.[52]

The battery industry has traditionally opposed the collection of batteries, whether for recycling or disposal as hazardous waste. NEMA has argued that collecting button cells in households increases their chance of being swallowed by children or the elderly. Furthermore, storing batteries in barrels creates a risk of fire or explosion from the hydrogen gas and the residual charge in the batteries. Finally, NEMA has pointed out that recycling programs, particularly those including alkaline batteries, may not be able to pay for themselves.[53]

Environmentalists claim that existing collection programs have proved feasible and that, by educating people to the dangers presented by batteries, household hazardous waste collection programs could prompt consumers to store them more carefully, thus decreasing the risk of accidental ingestion. Using vented containers in collection programs may alleviate the danger of explosions by preventing a buildup of hydrogen gas, and it might be possible to design packages for new button cells that could be reused to store used cells until they could be collected.

NEMA recently stated that any battery recycling programs to recover mercury should ideally be restricted to mercury-oxide batteries.[54] Since consumers cannot generally distinguish between different types of button cells, NEMA feels that recycling programs may need to collect all types of button cells. Interestingly, while at least one manufacturer labels its packaging with the composition of the battery (mercury, zinc, lithium, etc), the batteries themselves are not labeled, making separation at the point of disposal virtually impossible. This is likely to change, as Minnesota's battery law requires button cells to be labeled by type.

The NEMA Battery Products Alliance has also stated its support for the recycling of nickel-cadmium batteries.[55] As previously noted, most of the NiCad batteries used by consumers are sealed into products, and consumers would have to break the product open to remove the batteries. In 1989 Connecticut passed Public Act 89-385, which will require cordless rechargeable consumer products to be designed to facilitate removal of NiCad batteries by July 1, 1993, unless safety considerations indicate otherwise. Furthermore, it requires the products or their packaging or the batteries themselves to be labeled and will require municipal governments to collect the batteries (although the batteries could be landfilled as hazardous waste as well as recycled). NEMA has stated that it supports the legislation, although it would like to see the requirement restricted to *new* products so that manufacturers do not have to redesign and retool for current models.

Minnesota's battery law requires that, starting July 1, 1993, rechargeable batteries built into appliances be accessible to consumers and that the products be labeled as containing rechargeable batteries. It does not contain the exemptions the Connecticut law does, although it does contain a provision to grant a two-year extension if a product cannot otherwise be redesigned, if the redesign would result in significant hazard to consumers, or if the type of electrode used in the battery would not present a hazard in the waste stream. The law also requires manufacturers of mercury-oxide, silver-oxide, NiCad, and lead-acid batteries purchased by a government agency or an industrial, commercial, or medical facility to arrange for reclamation of the batteries or to take the batteries back themselves.

The future of battery recycling is uncertain. As there is a shift toward substitutes for mercury-oxides and NiCads, it will become less economical to recycle batteries, and existing collection programs may be abandoned. There are also questions about how lithium, nickel hydride, and mercury substitutes will affect the technical operation of recycling programs.

If adopted, a recent proposal by EPA could increase the separation and collection of batteries, if not their recycling. EPA proposed air emission standards for municipal waste incinerators in November 1989 with a 25 percent materials separation requirement. This controversial plan includes a proposal to require that batteries be separated from the solid waste stream. Such a plan could make economic sense if removing batteries allowed incinerator ash to be disposed of as nonhazardous waste instead of hazardous waste (since the former is significantly cheaper than the latter).[56]

Recycling is not a benign activity. It results in air and effluent emissions and sludge disposal. Some facilities have had difficulty meeting their permitted emissions limits. And no recycling program will collect 100 percent of the batteries sold. For instance, 73 percent of Japanese towns and cities, representing 72 percent of the population, collect cylindrical batteries for recycling or disposal, but only 9 percent of batteries sold are turned in. Retail outlets manage to collect just 27 percent of button cells.[57]

Recycling has a role for large industrial, medical, and military batteries. It probably has a role for certain consumer batteries (namely, mercury oxides and NiCads).[58] However, source reduction seems to be the most appropriate long-term solution for controlling the environmental impacts of household batteries, particularly the popular alkaline battery.

PART B. INFORMATION FOR SELECTING AN OPTION

If this were an actual application of the framework, it would be necessary to work through Part B of Tool 3 for each of the options selected in Step 2. Although the matrix at the end of the tool (figure B.15) shows the results for all of the options, for the sake of brevity, the section that follows demonstrates Part B for just two options: #11–"Rechargeables" and #13–"Use more efficiently." This is for the purpose of illustration only and is not meant as an endorsement of either of these two options.

PART B. OPTION 11–USE RECHARGEABLES

1. How This Option Applies to the Product and Its Potential Effectiveness for Reducing the Problem Identified

This option would replace primary batteries with NiCads, since they dominate the rechargeable household battery market. Using rechargeables instead of primary batteries could theoretically have a sizable impact on total quantity of batteries entering the waste stream, since one NiCad can replace 200 to 300 primary batteries. Figure B.14, from a study out of Carnegie-Mellon University, shows the results of a scenario based on using a cassette player two hours a day over a three-year period. Switching to NiCads could reduce the number of batteries used from 876 to 4.

The effect on toxicity, which is the real concern, is unclear. First, insufficient data is available to perform a comparative risk assessment. Second, the table assumes that the mercury content of the alkalines is 0.44 percent. Given manufacturers efforts at mercury reduction, this may be an overestimate in the future.

2. Other Effects

a. Environmental Trade-offs

What follows is a very rough qualitative assessment of key trade-offs during the life cycle of alternative battery types. A much more rigorous assessment would be required for any definite conclusions to be drawn.

Raw Materials Production

Quite a bit of information would have to be collected before one could make any comparison of the environmental effects of producing nickel, cadmium, manganese dioxide, zinc, carbon, mercury, or any of the other components. Note, however, that cadmium is a byproduct of zinc production. Depending on the level of zinc production for other uses, this could mean that there is relatively little marginal environmental burden from cadmium production.

Energy Consumed in the Product's Manufacture

No information was identified on this topic.

Materials Consumed Manufacturing the Product

Rechargeables are clearly superior from a materials conservation standpoint, since one NiCad is the functional replacement of 200-300 primary batteries.

Wastes Produced during Manufacture

Because of the paucity of information on this subject, no comparison was made. Even if information on quantities of waste were available from a life-cycle inventory, it would be difficult to compare thc environmental effects. What follows are some general observations on manufacturing wastes.

Solid Waste. According to a materials balance prepared for EPA in 1975, "The major loss of mercury in the manufacturing process occurs when batteries are rejected during inspection, testing, and packaging."[59] Plant scrap has been estimated at 1 to 5 percent of production and is probably at the low end of that range. Given that there are over three billion batteries sold a year, this represents a sizable number of batteries. Depending on the manufacturer, scrap alkaline batteries may be sent to hazardous waste landfills or local municipal landfills, or disposed of on site.

Plant scrap represents a small share of the industry's mercury consumption. The 1975 report for EPA went on to note that 95 percent of the mercury lost to the environment from the battery manufacturing and use cycle occurred when consumers disposed of their batteries. Unlike plant scrap, these batteries are widely dispersed throughout the nation's municipal waste stream.

A 1989 report for EPA assumed that manufacturing losses for NiCads were 15 percent.[60] Because of the high value of nickel and cadmium, NiCad manufacturers do recycle their plant scrap.

Water Emissions. According to EPA's 1984 Development Document for Effluent Limitations Guidelines and Standards for Battery Manufacturing:

> The most important pollutant parameters generated in battery manufacturing wastewaters are (1) toxic metals—arsenic, cadmium, chromium, copper, lead, mer-

Figure B.14
Theoretical Model: Batteries Saved by Switching a Cassette Player from Alkaline to NiCad Batteries*

Model factors	Alkaline scenario	NiCadscenario
Battery life span	5 hr	2 hr
Number of batteries used	876	4
Cost of batteries used	$657	$11
Number of recharges	—	1,095
Cost of battery recharger	—	$10
Cost of electricity for recharging time	—	$5
Total cost of batteries, recharger, and electricity	$657	$26
Wastes generated		
Mass of batteries used	24,100 grams	90 grams
Mass of heavy metals		
Mercury	105 grams	—
Cadmium	—	16 grams

*Theoretical results of a scenario based on using a cassette player two hours a day over a three-year period.

Source: Carnegie-Mellon University.

> cury, nickel, selenium, silver, and zinc; (2) nonconventional pollutants—aluminum, cobalt, iron, manganese, and COD [chemical oxygen demand]; and (3) conventional pollutants—oil and grease, TSS [total suspended solids], and pH. Toxic organic pollutants generally were not found in large quantities although some cyanide was found in a few subcategories. Because of the amount of toxic metals present, the sludges generated during wastewater treatment generally contain substantial amounts of toxic materials.[61]

The effluent guidelines were implemented in 1987. No material was identified on how successful plants have been at meeting their limits. Information in the Toxics Release Inventory (collected under Section 313 of the Emergency Planning and Community Right-to-Know Act of 1986) could possibly provide some information on total releases to the environment.

Air Emissions. No information was identified. Again, the Toxics Release Inventory might be a source of information.

Are Raw or Recycled Materials Used in Manufacturing?

Given the low level of recycling of batteries that currently occurs, it is doubtful there is significant use of recycled material. Only 4 percent of the mercury consumed in the United States in all uses is from recycled sources.[62] NiCads have a better potential for using recycled material, since cadmium and nickel can be economically reclaimed from batteries.

Occupational Exposure from the Production Process

No information was identified.

Product Packaging

Batteries are generally packaged in paperboard and plastic. Primary and rechargeable batteries have the same type of packaging. Rechargeables thus have a clear benefit in terms of total volume of packaging, since one rechargeable can replace 200 to 300 primary batteries.

Energy Consumed in Transporting the Product to Market

NiCads have a clear advantage, since one can replace 200 to 300 primary cells.

Occupational Exposure from Use

Not relevant to this category.

Energy Consumed in Use

Recharging of NiCads is very efficient. Primary cells, of course, are not recharged in use.

Waste Produced during Use

Since NiCads must be recharged, they indirectly contribute to waste production at electric power plants.

Available Disposal Options

Most batteries are disposed into the solid waste stream, where they are landfilled or incinerated. There is currently no domestic recycling of cylindrical cells. NiCads appear to have more potential to be recycled.

Environmental Effects of Recycling

Recycling conserves natural resources and removes material from the waste stream. However, if improperly operated, recycling facilities can also create a threat to the environment. No material was identified comparing the effects of mercury and cadmium recycling.

Environmental Effects of Disposal

Both mercury and cadmium may present risks if landfilled or incinerated. Incineration is generally agreed to be the predominant risk. Because of its volatility, mercury is more likely to escape in flue gas. Since cadmium is captured in incinerator ash, exposure depends on how the ash is handled. Data to determine whether disposal of one NiCad battery, at 18 to 22 percent cadmium, provides a risk reduction compared to disposal of 200 to 300 primary batteries, each with a fraction of a percent of mercury, was not identified.

Overall Assessment of Environmental Effects

The evidence is insufficient at this point to determine whether the environmental effects of the production, use, and disposal of NiCads are greater or less than the effects of alkalines.

b. Effect of This Option on Product Performance

One of the key performance characteristics for batteries is their length of life. For rechargeable batteries, this is both the overall life and the cycle-life (how often they must be recharged). Although NiCads may be recharged up to 1,000 times (lasting several years in some applications), their cycle-life is significantly shorter than the lifespan of an alkaline battery. The Carnegie-Mellon study assumed that the NiCads would need to be recharged after two hours, while the alkalines would last five hours.

In terms of other measures of performance, some products may not operate well on NiCads because of their current requirements, frequency of use, etc

c. Effect of This Option on Product Price and Sales

NiCads are more expensive than primary batteries and require the capital investment of a recharger.* But they can often save consumers money, particularly in products that receive frequent use. The Carnegie-Mellon scenario estimated the total cost of NiCads at $26, as compared to $657 for the equivalent number of alkalines.

If consumers could be induced to buy rechargeables, the result would be to decrease both the value and quantity of battery sales.

d. Effect of This Option on Manufacturers, Retailers, and Distributors

Were this option 100 percent effective (that is, a total switch in household cylindrical batteries from primaries to rechargeables), it would have a dramatic impact on manufacturers and retailers.† There would be a significant decrease in the dollar value of annual battery sales. In the short run, there would be a shift among producers, since most of the rechargeable batteries sold are not manufactured by the three industry leaders (Duracell, Eveready, and Rayovac). Presumably they could switch to producing rechargeables in the long run.

The decrease in sales volume would also have an impact on retailers. Retailers find batteries to be very attractive, since they take up little shelf space, have a high sales volume, and have a 30 to 40 percent profit margin.[63] The decrease in battery sales that would occur with a total switch to rechargeables would have a negative impact on retailers and distributors. But it is unlikely the effect would be large enough to create problems for retailers.

e. Effect of This Option on Complementary Products

Switching to NiCads would cause an increase in sales of battery rechargers. Thus, it might make sense to combine this option with an option to promote the use of efficient rechargers.

*For example, one drug store in Washington, D.C., was selling double-packs of D-cells at the following prices: $1.99 for carbon-zinc, $2.19 for house-brand alkalines, $3.99 for name-brand alkalines, and $6.49 for rechargeables (with the recharger selling for $12.49.

†Given consumer preference obstacles, it is unlikely that implementation of this option could be very effective. However, that question is addressed in Tool 5. Tool 3 assumes complete effectiveness of implementation.

f. Effect of This Option on Alternative Products

No effect identified.

g. Effect of This Option on Recycling and Other Waste Management Options

NiCads have a better potential for recycling. The environmental impacts would depend on how well recycling facilities controlled their emissions and how they handled their residual wastes.

3. Technical Barriers to This Option

None identified.

4. Other

None identified.

PART B. OPTION 13—USE MORE EFFICIENTLY

1. How This Option Applies to the Product and Its Potential Effectiveness for Reducing the Problem Identified

Purchasing the proper batteries for an application, and using them correctly, can improve their performance. Recommendations by manufacturers on choosing batteries include the following:

- Alkalines are best suited for products with high drain rates, such as stereo cassette players, camera flashes, and toys.
- Alkalines also work well in low or medium drain appliances such as flashlights, clocks, and radios. Carbon-zincs may also work well and are less expensive.
- Rechargeables are best suited to appliances that are used frequently, because that is when they are most economical.

Also, see the section on "Consumer misuse" in Part A.3 of Tool 3 for further discussion of recommended practices for using batteries.

If implemented successfully, this option would have a relatively small effect on the quantity of batteries in the municipal solid waste stream (and thus the toxicity of solid waste), since inefficient use does not have a major effect on the number of batteries consumed.

2. Other Effects

a. Environmental Trade-offs

None identified.

b. Effect of This Option on Product Performance

This option would improve product performance, by definition.

c. Effect of This Option on Product Price and Sales

Consumers can save money by purchasing the battery that is best suited to a particular product and then using it carefully. This would cause a minor decrease in the number of batteries sold.

d. Effect of This Option on Manufacturers, Retailers, and distributors

Reduction in sales could theoretically create a negative impact on manufacturers, retailers, and distributers, but the effect would not be large enough to have meaningful consequences.

e. Effect of This Option on Complementary Products

None identified.

f. Effect of This Option on Alternative Products

None identified.

g. Effect of This Option on Recycling and Other Waste Management Options

None identified.

3. Technical Barriers to This Option

None identified.

* * *

At this point in an actual application of the framework, it would be necessary to select from among the options previously identified. For the purpose of this illustration, Option 13–"Use more efficiently" will be used to demonstrate the remainder of the framework. This is not intended as an endorsement of this option over any other option.

Figure B.15
Sample Matrix for Selecting Promising Source ReductionOptions

Summary of analysis for product of concern / Options →	Eliminate product: Don't buy	Eliminate toxics: Reduce mercury / cadmium	Increase life span / Improve efficiency	Resue: Rechargeables	Efficient use	Complementary products: Battery-driven products	Complementary products: Rechargers
1. *Effectiveness of source reduction option*							
a. Overall effectiveness in solving primary problem (in this case, toxicity)	++	+?	+?	?	+?	+	+
2. *Other effects from implementing the option*							
a. Environmental trade-offs	+?	?	?	?	/	?	?
b. On product performance	--	?	++	-	+	+	+
c. On product price and sales	↓ Sales	?	↓ Sales	↓ Sales ↓ Price (for some)	↓Sales	↓ Sales	↓ Sales
d. On manufacturers, retailers, and distributors	--	/	-?	-?	-?	-?	-?
e. On complementary products	↑ Nonbattery products	/	/	↑ Rechargers	/	++	/
f. On alternative products	↑ Residential electricity use	/	/	↑ Residential electricity use	/	/	/
g. On recycling and other waste management options	/	-?	/	+	/	/	/
3. *Technical barriers to implementing the option*	/	?	?	/	/	?	/
4. *Other*							

For options scanned in Step 2, give a value of **?** for unclear, **--** for very negative, **-** for negative, **-?** for potentially negative, **+?** for potentially positive, **+** for positive, **++** for very positive, and **/** for no effect.

Tool 4.A
Obstacles to Source Reduction Options

There may be an information obstacle, since many consumers may not be aware of how to choose and use batteries efficiently.

Tool 4.B
Checklist and Guide to Source Reduction Strategies

Two strategies seem appropriate to provide the information consumers lack on using batteries efficiently (thus overcoming the obstacle to this option). They are:

- *Media/public outreach.* Newspaper and magazine ads would probably be most effective, because consumers could cut them out and save the information.
- *In-store shopper awareness campaign.* This could be accomplished though brochures or signs at sales racks or with an insert in battery packages.

It would be reasonable to combinc these two strategies into a single strategy package, focused on public information.

Tool 5
Evaluation Questions for Selecting Strategies

1. How Does This Strategy Apply to the Option Identified and Its Potential Effectiveness for Overcoming the Obstacles Identified?

Consumers can be provided with information about battery purchase and use that will allow them to use the batteries more efficiently, thus reducing waste. Even if consumers used batteries efficiently, this would have a small impact on the number of batteries sold. However, it is unlikely that this strategy package would be completely effective. First, a media campaign might not reach all consumers. Second, this would be somewhat difficult information to apply. Consumers might not know whether a particular appliance had a high drain rate or not or whether they used a product intensively enough to warrant purchasing rechargeable batteries.

2. How Feasible Is Implementation of This Strategy?

a. Who Are the Actors to Implement It?

The battery manufacturers, NEMA, and consumer groups would be good candidates to carry out a public information campaign. In fact, they already provide such information. Some battery packages are labeled with the type of appliance the battery works best in. Rayovac publishes a brochure entitled "A Consumer's Guide to Smart Battery Buying," which customers can write for, and advertises the brochure on its battery packages. And the periodic articles in *Consumer Reports* rating different brands of batteries provide tips on buying and using them. Finally, stores that sell batteries could provide information, as they do on the nutritional aspects of some foods.

b. What Resources Are Necessary for Implementation?

The primary resource needed would be money. How much money would depend on how large a program is envisioned. Given the amount of money that is spent on advertising for batteries, the industry apparently has the money available for some version of a public information campaign.

c. How Accessible Is the Information Required for Implementation?

All the information necessary to carry out this strategy is readily available.

3. What Would Be This Strategy's Burden?

a. Does This Strategy Involve the Least Amount of Intervention Necessary to Accomplish the Task?

This would be a voluntary strategy to provide information.

b. Is This Strategy Socially and Economically Equitable?

This strategy would not have any adverse impacts on social and economic equity.

c. What Is the Economic Efficiency of This Strategy as Applied to the Option under Consideration

This would depend on what types of media campaigns were used to disseminate this information. An extensive campaign that would inform and change the behavior of most or all of the population would be expensive. Given the small reduction in battery usage (and thus toxicity) that would be expected if this strategy were 100 percent effective, the costs might well outweigh the benefits. A more modest campaign, aimed at changing the behavior of only part of the population, might have a positive benefit-cost ratio but would provide even more modest gains. It is unclear whether it would be worthwhile to expand public information efforts beyond what is currently ongoing.

* * *

Figure B.16 (next page) displays a sample of how a matrix for selecting strategy packages might be filled out. In an actual application of the framework, one would need to decide whether to recommend that this strategy be implemented. No recommendation is made here, because that is outside the scope of this paper as an explanatory exercise.

Figure B.16
Summary Matrix for Selecting Implementation Strategies

Summary of analysis for product of concern	**Strategy Package 1** — Media / public outreach — In-store shopper awareness
1. Effectiveness of strategy	
a. Overall effectiveness in overcoming obstacles	+
2. Feasibility of implementation	
a. Actors to implement	
b. Resources	+
c. Access to information	++
3. Burden associated with the strategy	
a. Least amount of intervention	++
b. Social and economic equity	++
c. Economic efficiency	?
4. Other	

For strategies scanned in Step 4, give a value of **?** for unclear, **--** for very negative, **-** for negative, **-?** for potentially negative, **+?** for potentially positive, **+** for positive, and **/** for no effect.

References

CHAPTER 1. THE EMERGENCE OF SOURCE REDUCTION

1. U.S. Environmental Protection Agency, Office of Solid Waste, *Characterization of Municipal Solid Waste in the United States: 1990 Update*, prepared by Franklin Associates Ltd., EPA/530-SW-90-042 (Washington, D.C.: U.S. Environmental Protection Agency, 1990).

2. U.S. Congress, Office of Technology Assessment, *Facing America's Trash: What Next for Municipal Solid Waste?*, OTA-O-424 (Washington, D.C.: U.S. Governmnet Printing Office, 1989).

3. Richard A. Denison and John Ruston, eds., *Recycling & Incineration: Evaluating the Choices* (Washington, D.C.: Island Press, 1990).

4. U.S. Office of Technology Assessment, *Facing America's Trash*; Denison and Ruston, *Recycling & Incineration.*

5. U.S. Office of Technology Assessment, *Facing America's Trash.*

6. Ibid., p. 87.

7. "Convenience Packaging Continues to Pile Up," *Wall Street Journal*, August 7, 1990.

8. Karen Brattesani, Research Innovations, "Seattle Solid Waste Utility 1990 Waste Reduction Survey," prepared for Seattle Solid Waste Utility, Department of Engineering, April 1990.

9. U.S. Environmental Protection Agency, Office of Solid Waste and Emergency Response, *Report to Congress: Solid Waste Disposal in the United States*, Vol. 1 EPA/530-SW-88-011, October 1988.

10. U.S. Office of Technology Assessment, *Facing America's Trash.*

11. National Governors' Association, Task Force on Solid Waste Management, *Curbing Waste in a Throwaway World* (Washington, D.C.: National Governors' Association, 1990).

12. U.S. Environmental Protection Agency, "The Solid Waste Dilemma: Solutions for the 90's," preliminary draft, August 1990.

13. U.S. Environmental Protection Agency, *Characterization of Municipal Solid Waste in the United States: 1990 Update*.

14. Denison and Ruston, *Recycling & Incineration.*

15. National League of Cities, *Municipal Incinerators: 50 Questions Every Local Government Should Ask* (Washington, D.C.: National League of Cities, 1988); cited in Denison and Ruston, *Recycling & Incineration.*

16. David Riggle and Jim Glenn, "The State of Garbage in America: Part 1," *Biocycle*, April 1991, pp. 34-38. (Figures based on surveys of state and local officials.)

17. Institute for Local Self-Reliance, *Beyond 40 Percent: Record-Setting Recycling and Composting Programs* (Washington, D.C.: Institute for Local Self-Reliance, August 1990).

18. U.S. Environmental Protection Agency, *The Solid Waste Dilemma: Solutions for the 90's.*

19. Denison and Ruston, *Recycling & Incineration.*

20. Readers are encouraged to consult other reports that explore such data in much greater detail. For example, U.S. Environmental Protection Agency, *Characterization of Municipal Solid Waste: 1990* contains, for the first time, estimates of volume for many categories of the waste stream. See chapter 2 for discussion of major sources.

21. The total U.S. population is projected to increase from 268.3 million in the year 2000 to 282.6 million in 2010—a 5.3 percent increase; per capita generation rates are projected to increase from 4.41 pounds per day in 2000 to 4.86 pounds per day in 2010—a 10.2 percent increase.U.S. Department of Commerce, Bureau of the Census, *Statistical Abstract of the United States, 1990* (Washington, D.C.: Govern-

ment Printing Office, 1990); U.S. Environmental Protection Agency, *Characterization of Municipal Solid Waste: 1990*.

22. Total disposable personal income in the U.S., in constant 1982 dollars, has increased from $1,668 million in 1970 to $2,800.5 in 1980—a 68 percent increase; per capita disposable income has increased 40 percent over the same period. U.S. Department of Commerce, Bureau of Economic Analysis, *Survey of Current Business*, Table 8.2, July 1990.

23. J. Bagby, Seattle Solid Waste Utility, personal communication, October 1990.

24. William Ratje and Barry Thompson, *The Milwaukee Garbage Project*, prepared for the Solid Waste Council of the Paper Industry (Washington, D.C.: Solid Waste Council, 1981).

25. U.S. Office of Technology Assessment, *Facing America's Trash*; Rathjue and Thompson, *The Milwaukee Garbage Project*; Brattesani, *1990 Waste Reduction Survey*.

26. J. Bagby, personal comnmunication, October 1990.

27. U.S. Office of Technology Assessment, *Facing America's Trash*.

28. Minnesota Pollution Control Agency, *State Solid Waste Policy Report—A Focus on Greater Minnesota* (St. Paul: Minnesota Pollution Control Agency, November 1988).

29. U.S. Office of Technology Assessment, *Facing America's Trash*.

30. Brattesani, *1990 Waste Reduction Survey*.

31. Blair T. Bower, "Economic, Engineering, and Polity Options for Waste Reduction," in *Waste Reduction: Research Needs in Applied Social Sciences* (Washington, D.C.: National Academy Press, 1990).

32. For example, U.S. Office of Technology Assessment, *Facing America's Trash*; The Conservation Foundation, *State of the Environment: A View toward the Nineties* (Washington, D.C.: The Conservation Foundation, 1987); U.S. Council on Environmental Quality, *Environmental Quality—Twentieth Annual Report* (Washington, D.C.: Council on Environmental Quality, 1990).

33. Karen Hurst and Paul Relis, Gildea Resource Center, *The Next Frontier: Solid Waste Source Reduction* (Santa Barbara: Community Environmental Council, 1988).

34. Society of Environmental Toxicology and Chemistry (SETAC), *A Technical Framework for Life-cycle Assessments* (Washington, D.C.: SETAC, January 1991).

35. Ibid.

36. Francis H. Irwin and Barbara Rodes, *Making Decisions on Cumulative Impacts: A Guide for Managers*, prepared for the Council on Environmental Quality (Washington, D.C.: World Wildlife Fund & The Conservation Foundation, in press).

37. U.S. Environmental Protection Agency, *Toxic Chemical Release Inventory Risk Screening Guide*, EPA 560/2-89-002 (Washington, D.C.: U.S. Environmental Protection Agency, 1989).

38. Society of Environmental Toxicology and Chemistry (SETAC), *A Technical Framework for Life-cycle Assessments*.

Figures

1.1. Adapted from U.S. Environmental Protection Agency, Office of Solid Waste, *Characterization of MunicipalSolid Waste in the United States: 1990 Update*, prepared by Franklin Associates Ltd., EPA/530-SW-90-042 (Washington, D.C.: U.S. Environmental Protection Agency, 1990).

1.2. Based on figures in ibid.; U.S. Environmental Protection Agency, "The Solid Waste Dilemma: Solutions for the 90's," preliminary draft, August 1990.

1.3. Adapted from U.S. Environmental Protection Agency, *Characterization of Municipal Solid Waste in the United States: 1990 Update*.

1.4. See U.S. Congress, Office of Technology Assessment, *Facing America's Trash: What Next for Municipal Solid Waste?*, OTA-O-424 (Washington, D.C.: U.S. Governmnet Printing Office, 1989) for discussion of routes to source reduction displayed in this figure. Also see chapter 2 of this report for a guide to these and other options.

1.7–1. Paul Kaldjian, personal communication, April 1991. While proposing a 10 percent reduction goal for the year 2000 in its August 1990 draft update of the *Agenda for Action*, EPA has tentatively decided not to establish a numeric goal at this time due to difficulties in measuring progress.

1.7–2. National Governors' Association, Task Force on Solid Waste Management, *Curbing Waste in a Throwaway World* (Washington, D.C.: National Governors' Association, 1990).

1.7–3. Coalition of Northeast Governors, *Source Reduction Council of CONEG: Progress Report* (Washington D.C.: CONEG Policy Research Center, Inc., 1990).

1.7–4. National Governors' Association, *Curbing Waste in a Throwaway World*.

1.7–5. Ed Fox and Tom Rattray, personal communication, October 1990.

CHAPTER 2: EVALUATING OPPORTUNITIES FOR SOURCE REDUCTION

1. U.S. Environmental Protection Agency, Office of Solid Waste, *Characterization of MunicipalSolid Waste in the United States: 1990 Update*, prepared by Franklin Associates Ltd., EPA/530-SW-90-042 (Washington, D.C.: U.S. Environmental Protection Agency, 1990).

2. U.S. Congress, Office of Technology Assessment, *Facing America's Trash: What Next for Municipal Solid Waste?*, OTA-O-424 (Washington, D.C.: U.S. Governmnet Printing Office, 1989); U.S. Environmental Protection Agecny, Office of Solid Waste, *The Solid Waste Dilemma: An Agenda for Action* (Washington, D.C.: U.S. Environmental Protection Agency, 1989).

3. U.S. Environmental Protection Agency, Office of Solid Waste, *Characterization of Products Containing Lead and Cadmium in Municipal Solid Waste in the United States, 1970 to 2000—A Final Report*, EPA 530-SW-89-015B (Washington, D.C.: U.S. Environmental Protection Agency, January 1989). See also U.S. Office of Technology Assessment, *Facing America's Trash*, for discussion on toxic constituents in the waste stream and research needs.

CHAPTER 3: SELECTED STRATEGIES TO ENCOURAGE SOURCE REDUCTION

1. U.S. Environmental Protection Agency, Office of Solid Waste, *Characterization of Municipal Solid Waste in the United States: 1990 Update*, prepared by Franklin Associates Ltd., EPA/530-SW-90-042 (Washington, D.C.: U.S. Environmental Protection Agency, 1990).

2. U.S. Environmental Protection Agency, *Yard Waste Composting: A Study of Eight Programs*, EPA/530-SW-89-038 (Washington, D.C.: U.S. Environmental Protection Agency, 1989).

3. Lisa Skumatz and Cabell Breckinridge, *Variable Rates in Solid Waste: Handbook for Local Solid Waste Officials* (Seattle: Seattle Solid Waste Utility, 1990).

4. Marvin Katz, "Collection Strategies of the Nineties," *Waste Age*, February 1989, pp. 60-68; David Riggle, "Only Pay for What You Throw Away," *BioCycle*, February 1989, pp. 39-41; Kimberly Sproule and Jeanne Cosulich, "Higher Recovery Rates: The Answer's in the Bag", *Resource Recycling*, November/December 1988, pp. 20, 43-44.

5. Katz, "Collection Strategies of the Nineties"; Riggle, "Only Pay for Why You Throw Away"; Sproule and Cosulich, "Higher Recovery Rates."

6. Pamela Winthrop Lauer and Neal Miller, "Minnesota County Participates in Waste Reduction Project," *Resource Recycling*, February 1990, p.50; League of Women Voters of Washington, *Solid Waste Reduction and Recycling: A Handbook of Strategies Employed by Businesses in Washington State*, prepared for the Washington State Department of Ecology, Office of Waste Reduction (Olympia, Wash.: Department of Ecology, 1989); Rhode Island Department of Environmental Management, *Handbook for Reduction and Recycling of Commercial Solid Waste* (Providence, R.I.: Department of Environmental Management, 1988).

7. Pennsylvania Resources Council, *Become an Environmental Shopper*, (Media, Pa.: Pennsylvania Resources Council, 1988).

8. Teresa Jones, et al., *Environmental Curricula Concerning Waste Management* (Amherst, Mass.: Northeast Regional Environmental Public Health Center, 1989).

9. U.S. Environmental Protection Agency, Office of Solid Waste and Emergency Response, *Let's Reduce and Recycle: Curriculum for Solid Waste Awareness*, EPA/530-SW-90-005 (Washington, D.C.: Environmental Protection Agency, June 1990).

10. Jones, *Environmental Curricula Concerning Waste Management*.

11. Ibid.

12. U.S. Congress, Office of Technology Assessment, *Facing America's Trash: What Next for Municipal Solid Waste?*, OTA-O-424 (Washington, D.C.: U.S. Governmnet Printing Office, 1989).

APPENDIX B: APPLICATION OF THE FRAMEWORK TO HOUSEHOLD BATTERIES

1. Amir Ghazall, "The Assault on Batteries", *Chemical Business*, July/August 1989, p.36.

2. "Innovative Battery Promotions Charge Drug Store Sales," *Drug Store News*, June 20, 1988, p.116; Kanner, Bernice, "Market for Batteries Is Now Supercharged," *Drug Store News*, February 1, 1988, p.26.

3. "Batteries Are Still Making Waves Through Niche Marketing," *Drug Store News*, January 8, 1990, p.21.

4. Information provided by the National Electrical Manufacturers Association.

5. Minnesota Pollution Control Agency, Karen Arnold, Nancy Misra, and Randall Hukriede, *Household Batteries in Minnesota: Interim Report of the Household Battery Recycling and Disposal Study (St.*

Paul: Minnesota Pollution Control Agency, 1990), appendix A.

6. U.S. Department of the Interior, Bureau of Mines, *Mercury Minerals Yearbook 1988* (Pittsburgh: Bureau of Mines, 1988), p.4.

7. Ibid.

8. Minnesota Pollution Control Agency, *Household Batteries in Minnesota*, appendix A.

9. Carnegie Mellon University, Department of Engineering & Public Policy, School of Urban and Public Affairs, and Department of Social and Decision Sciences, *Household Batteries: Is There a Need for Change in Regulation and Disposal Procedure?* (Pittsburgh: Carnegie Mellon University, 1989), p.18.

10. David Linden, ed., *Handbook of Batteries and Fuel Cells* (New York: McGraw-Hill, 1984), pp. 8-25.

11. Information provided by the National Elcctrical Manufacturers Association.

12. Ibid.

13. Linden, *Handbook of Batteries and Fuel Cells*, p. 18-1.

14. Minnesota Pollution Control Agency, Division of Water Quality, *Assessment of Mercury Contamination in Selected Minnesota Lakes and Streams: Report to the Legislative Commission on Minnesota Resources*, Executive Summary (St. Paul: Minnesota Pollution Control Agency, 1989), p.3.

15. Minnesota Pollution Control Agency, *Household Batteries in Minnesota*, p.20.

16. Ibid., pp.47-60.

17. Ibid., p.20.

18. Ibid., p.42.

19. Carnegie Mellon University, Household Batteriesy p.113.

20. Minnesota Pollution Control Agency, *Household Batteries in Minnesota*, p.28.

21. "Dry-cell Batteries," *Consumer Reports*, November 1987, p.705.

22. Franklin Associates Ltd., *Characterization of Products Containing Lead and Cadmium in Municipal Waste in the United States, 1970 to 2000*, prepared for U.S. Environmental Protection Agency, Office of Solid Waste, EPA/530-SW-89-015A (Washington, D.C.: U.S. Environmental Protection Agency, January 1989), p.157.

23. Minnesota Pollution Control Agency, *Household Batteries in Minnesota*, p.44.

24. Science Application Inc., *Economic Impact Analysis of Effluent Limitations and Standards for the Battery Manufacturing Industry*, EPA 440/2-84-003 (Washington, D.C.: Environmental Protection Agency, February 1984), prepared for U.S. EPA, Office of Analysis and Evaluation, p.4-12; Linden, *Handbook of Batteries and Fuel Cells*, p.9-1.

25. Trudy Bell, "Choosing the Best Battery for Portable Equipment," *IEEE Spectrum*, March 1988, pp.31-33.

26. Timothy Somheli, "Charging the Industry," *Appliance*, February 1990, p.49; Douglas Bahniuk, "New Lithium Cells Charge Up Consumer Electronics," *Machine Design, February 21, 1989, p.104; Dana Gardner, "New Look for Lithium Batteries," Design News*, November 6, 1989, p.97.

27. David Cohen, "Environmentally Sound Disposition of Household Batteries," presented to the Legislative Commission on Solid Waste Management Conference on Solid Waste Management and Materials Policy, New York, N.Y., February 2, 1990, p.3; Carnegie Mellon, *Household Batteries*, p.86.

28. "Hydride Batteries Charge Up", *Chemical Week*, May 25, 1988, p.9.

29. Minnesota Pollution Control Agency, *Household Batteries in Minnesota*, p.44.

30. Clare Ansberry, "Eastman Kodak Is Pulling Plug on Its Ultralife," *Wall Street Journal*, April 10, 1990, p.B-l.

31. Barbara Berkman, "Lithium Batteries Rewrite the Rules for Laptop Makers," *Electronic Business*, February 6, 1989, p.76.

32. Ghazall, "The Assault on Batteries," p.36.

33. Personal communication with Terry Telzrow, Eveready, June 1990.

34. Information provided by the National Electrical Manufacturers Association.

35. Barbara Berkman, "Lithium Batteries," p.76.

36. Timothy Somheli, "Chargin the Industry," p.51.

37. Personal communication with Ray Balfour, Rayovac Corporation, March 1990.

38. "Dead Batteries No Longer Need to Leak Mercury," *New Scientist*, October 7, 1989, p.36.

39. Personal communication with Tay Balfour, Rayovac Corporation, March 1990.

40. Timothy Forker, "Strategic Approaches to the Used Household Battery Problem; A Report on European Experiences and Their Implications for Action in the United States," Environmental Action Coalition, New York, NY., November 10, 1989; Timothy Forker, "Household Batteries: Recycling & Toxicity Reduction in Europe," presented at the New York State Legislative Commission on Solid Waste Management's Sixth Annual Conference on Solid Waste Management and Matcrials Policy, New York City, February 2, 1990; "EC Ministers Prohibit Batteries Containing Certain Dangerous Substances,"

Bureau of National Affairs, *International Environment Reporter*, June 1990, p.223; Allen Hershkowitz and Mary Uva, "Waste Reduction Has A Way To Go," *Waste Age*, September 1990, pp.62-64.

41. Terry Telzrow, "Mercury Reduction: A Success Story in the Battery Industry", *Proceedings of the First International Seminar on Battery Waste Management*, Deerfield Beach, November 6-8, 1989; cited in Minnesota Pollution Control Agency, *Household Batteries in Minnesota*, p.17.

42. "Performance, Price Charge Battery Buying," *Drug Store News*, December 11, 1989, p.25; Bernice Kanner, "Market for Batteries," p.26.

43. U.S. EPA, Effluent Guidelines Division, *Development Document for Effluent Limitations Guidelines and Standards for Battery Manufacturing*, EPA 440/1-84/067 (Washington, D.C.: U.S. Environmental Protection Agency, 1984), p. 83.

44. *Drug Store News*, January 8, 1990, p.21.

45. *Drug Store News*, February 1, 1988, p.27.

46. "A Consumer Guide to Smart Battery Buying," Rayovac Corporation.

47. Carnegie-Mellon University, *Household Batteries*, p.86.

48. Vickie Schlierer, "Current Technology of Battery Recycling and Disposition," paper presented to the EPA Task Force on Precombustion Control of Mercury Emissions from Batteries, Research Triangle Park, North Carolina, February 8, 1990.

49. Personal communication with Mike Roller, Recytec America, INc., March 1990.

50. Personal communication with Hugh Morrow, The Cadmium Council, October 1990.

51. Personal communication with John Onuska, Jr., INMETCO (International Metal Reclamation Company, Inc.), April 1990.

52. Personal communication with Fred Wehmeyer, Gates Energy Products, March 1990.

53. National Electrical Manufacturers Association, *Written Statement of the National Electrical Manufacturers Association and the Battery Products Alliance Concerning Household Battery Disposal* (Washington, D.C.: National Electrical Manufacturers Association, 1988).

54. National Electrical Manufacturers Association, "Household and Other Batteries: Source Reduction and Recycling," Presented at the 9th Annual Resource Recovery Conference of the U.S. Conference of Mayors, Washington, D.C., March 30, 1990 by Raymond L. Balfour on behalf of the Dry Battery Section, National Electrical Manufacturers Association and the Battery Products Alliance.

55. Ibid.

56. Tom Watson, "EPA's Recycling Proposal: The Waste Hits the Fan," *Resource Recycling*, February 1990, p.78.

57. Lord Deirdre, "Burnt Out Batteries", *Environmental Action*, September/October 1988, p.18.

58. Randy Johnson and Carl Hirth, "Collecting Household Batteries," *Waste Age*, June 1990, p.49.

59. URS Research Company, "Materials Balance and Technology Assessment of Mercury and its Compounds on National and REgional Bases," prepared for U.S. Environmental Protection Agency, Office of Toxic Substances, EPA-560/3-75-007, October 1975, p.174.

60. Franklin Associates, *Characterization of Products*, p.157.

61. U.S. EPA, Effluent Guidelines Division, "Development Document for Effluent Limitations Guidelines," p.2.

62. Signal Environmental Systems, "Source of Heavy Metals in Municipal Solid Waste," Des Plaines, IL, Oct. 3, 1986, prepared for NH/VT Solid Waste Project, p.2.

63. Bernice Kanner, "Market for Batteries," p.26.

Bibliography

GENERAL

Applied Decision Analysis, Inc. *Environmental Labeling in the United States: Background Research, Issues, and Recommendations*. Prepared for U.S. Environmental Protection Agency, Office of Pollution Prevention. Menlo Park, Calif.: Applied Decision Analysis. 1990.

Brattesani, Karen, Research Innovations. "Seattle Solid Waste Utility 1990 Waste Reduction Survey." Prepared for Seattle Solid Waste Utility, Engineering Department." April 1990.

Coalition for Northeast Governors. *Final Report of the Source Reduction Task Force*. Washington, D.C.: Policy Research Center, Inc. 1989.

Coalition for Northeast Governors. Source Reduction Council of CONEG. *First Annual Report*. Washington, D.C.: Policy Research Center, Inc. 1990.

Council for Solid Waste Solutions. *The Solid Waste Management Problem: No Single Cause, No Single Solution*. Washington, D.C.: Council for Solid Waste Solutions. 1989.

Denison, Richard A., and John Ruston, eds.. *Recycling and Incineration: Evaluating the Choices*. Washington, D.C.: Island Press. 1990.

Environmental Data Services, Ltd. *Eco-Labels: Product Management in a Greener Europe*. Camberley, England: Southwell Press Ltd. undated.

Hurst, Karen, and Paul Relis, Gildea Resource Center. *The Next Frontier: Solid Waste Source Reduction*. Santa Barbara, Calif.: Community Environmental Council. 1988.

Matrix Management Group. *Waste Stream Composition Study, 1988-1989*. Final Report. Prepared for the City of Seattle, Department of Engineering, Solid Waste Utility. 1989.

National Governors' Association. *Curbing Waste in a Throwaway World*. Report of the Task Force on Solid Waste Management. Washington, D.C.: National Governors' Association. 1990.

National Research Council. *Waste Reduction: Research Needs in Applied Social Sciences*. Washington, D.C.: National Research Council. 1990.

Rathje, William, V.W. Lambov, and R.C. Herndon, University of Arizona Bureau of Applied Research in Anthropology, and Florida State University, Center for Biomedical and Toxicological Research and Hazardous Waste Management. *Characterization of Household Hazardous Waste from Marin County, California, and New Orleans, Louisiana*. Prepared for the U.S. Environmental Protection Agency. EPA/600/4-87/025. Las Vegas, Nev.: U.S. Environmental Protection Agency. 1987.

Rathje, William, Michael Reilly, and Wilson Hughes, University of Arizona Bureau of Applied Research in Anthropology. *Household Garbage and the Role of Packaging: The United States/Mexico City Household Refuse Comparison*. Prepared for the Solid Waste Council of the Paper Industry. Washington, D.C.: The Solid Waste Council of the Paper Industry. 1985.

Rathje, William. "Rubbish." *The Atlantic Monthly*. December 1989. Pp. 99-109.

Resource Integration Systems, Ltd., in association with Victor and Burrell Research and Consulting. *Barriers to Reduction, Recycling, Exchange and Recovery of Special Waste in Ontario*. Prepared for Ontario Waste Management Corporation. Toronto: Ontario Waste Management Corporation. 1984.

Rhode Island Solid Waste Management Corporation. *Rhode Island Solid Waste Composition Study*. Providence, R.I.: Rhode Island Solid Waste Management Corporation. 1990.

Selke, Susan E. M. *Packaging and the Environment: Alternatives, Trends and Solutions*. Lancaster, Pa.: Technomic Publishing Company. 1990.

U.S. Congress. Office of Technology Assessment. *Facing America's Trash: What Next for Municipal*

Solid Waste?. OTA-O-424. Washington, D.C.: U.S. Government Printing Office. 1989.

U.S. Environmental Protection Agency. *Toxic Chemical Release Inventory Risk Screening Guide*. EPA 560/2-89-002. Washington, D.C.: U.S. Environmental Protection Agency. 1989.

_________. Office of Solid Waste. *The Solid Waste Dilemma: An Agenda for Action*. EPA/530-SW-89-019. Washington D.C.: U.S. Environmental Protection Agency. 1989.

_________. _________. *Characterization of Municipal Solid Waste in the United States: 1990 Update*. Prepared by Franklin Associates Ltd. EPA/530-SW-90-042. Washington, D.C.: U.S. Environmental Protection Agency. 1990.

_________. _________. *The Solid Waste Dilemma: An Agenda for Action, Background Document*. EPA/530-SW-88-054B. Washington, D.C.: U.S. Environmental Protection Agency 1988.

_________. _________. "The Solid Waste Dilemma: Solutions for the 90's." Preliminary draft. August 1990.

_________. Office of Solid Waste and Emergency Response / Office of Water. *Methods to Manage and Control Plastic Wastes*. EPA/530-SW-89-051. Washington, D.C.: U.S. Environmental Protection Agency. 1990.

_________. _________. Policy, Planning and Evaluation. *Promoting Source Reduction and Recyclability in the Martetplace*. EPA 530-SW-89-066. Washington, D.C.: Environmental Protection Agency. 1989.

_________. Science Advisory Board. *Reducing Risk: Setting Priorities and Strategies for Environmental Protection*. SAB-EC-90-021. Washington, D.C.: U.S. Environmental Protection Agency, 1990.

Young, John E. *Discarding the Throwaway Society*. Worldwatch Paper 101. Washington, D.C.: Worldwatch Institute. 1991.

Zimmerman, Elliot. *Solid Waste Management Alternatives: Review of Policy Options to Encourage Waste Reduction*. ILENR/RE-PA-88/03. Springfield, Ill.: Illinois Department of Energy and Natural Resources, Research and Planning Section. 1988.

THINGS THAT CAN BE DONE NOW

Jones, Teresa, et al.. *Environmental Curricula Concerning Waste Management*. Amherst, Mass.: Northeast Regional Environmental Public Health Center. 1989.

Katz, Marvin. "Collection Strategies of the Nineties." *Waste Age*. February 1989. Pp. 60-68.

Lauer, Pamela Winthrop, and Neal Miller. "Minnesota County Participates in Waste Reduction Project." *Resource Recycling*. February 1990. P. 50.

League of Women Voters of Washington. *Solid Waste Reduction and Recycling: A Handbook of Strategies Employed by Businesses in Washington State*. Prepared for the Washington State Department of Ecology, Office of Waste Reduction. Olympia, Wash.: Department of Ecology. 1989.

Pennsylvania Resources Council. *Become an Environmental Shopper*. Media, Pa.: Pennsylvania Resources Council. 1988.

Rhode Island Department of Environmental Management. *Handbook for Reduction and Recycling of Commercial Solid Waste*. Providence, R.I.: Department of Environmental Management. 1988.

Rhode Island Department of Environmental Management and Rhode Island Solid Waste Management Corp. *Rhode Island Source Reduction Task Force Report*. Providence, R.I.: Rhode Island Waste Management Corporation. 1987.

Rhode Island Solid Waste Managment Corporation. "Don't Let Your Dollars/Bottom-line Go to Waste." Poster. October 1988.

Riggle, David. "Only Pay for What You Throw Away." *BioCycle*. February 1989. Pp. 39-41.

Skumatz, Lisa, and Cabell Breckinridge. *Variable Rates in Solid Waste: Handbook for Local Solid Waste Officials*. Seattle: Seattle Solid Waste Utility. 1990.

Sproule, Kimberly, and Jeanne Cosulich. "Higher Recovery Rates: The Answer's in the Bag." *Resource Recycling*. November/December 1988. Pp. 20, 43-44.

U.S. Environmental Protection Agency. *Yard Waste Composting: A Study of Eight Programs*. EPA/530-SW-89-038. Washington, D.C.: U.S. Environmental Protection Agency. 1989.

_________. Office of Solid Waste and Emergency Response. *Let's Reduce and Recycle: Curriculum for Solid Waste Awareness*. EPA/530-SW-90-005. Washington, D.C.: U.S. Environmental Protection Agency. 1990.

Watson, Tom. "New Yard Waste Strategy: Pay-by-the-Bag." *Resource Recycling*. February 1990. Pp. 35-37.

Woestendiek, Carl, and Susan J. Smith. "Implementing a Backyard Composting Program." *Biocycle*. December 1990. Pp. 70-71.

Index

Also Available from Island Press

Business and the Environment: A Resource Guide
Management Institute for Environment and Business

Coastal Alert: Ecosystems, Energy, and Offshore Oil Drilling
By Dwight Holing

The Complete Guide to Environmental Careers
The CEIP Fund

Death in the Marsh
By Tom Harris

Energy and the Ecological Economics of Sustainability
By John Peet

The Global Citizen
By Donella Meadows

Healthy Homes, Healthy Kids
By Joyce Schoemaker and Charity Vitale

Inside the Environmental Movement: Meeting the Leadership Challenge
By Donald Snow

Last Animals at the Zoo: How Mass Extinction Can Be Stopped
By Colin Tudge

Learning to Listen to the Land
Edited by Bill Willers

Lessons from Nature: Learning to Live Sustainably on the Earth
By Daniel D. Chiras

The Living Ocean: Understanding and Protecting Marine Biodiversity
By Boyce Thorne-Miller and John G. Catena

Making Things Happen
By Joan Wolfe

Media and the Environment
Edited by Craig LaMay and Everette E. Dennis

Nature Tourism: Managing for the Environment
Edited by Tensie Whelan

Our Country, The Planet: Forging a Partnership for Survival
By Shridath S. Ramphal

Overtapped Oasis: Reform or Revolution for Western Water
By Marc Reisner and Sarah Bates

Plastics: America's Packaging Dilemma
By Nancy Wolf and Ellen Feldman

Rain Forest in Your Kitchen: The Hidden Connection Between Extinction and Your Supermarket
By Martin Teitel

The Rising Tide: Global Warming and World Sea Levels
By Lynne T. Edgerton

The Rolling Stone Environmental Reader
Rolling Stone magazine

Steady-State Economics: Second Edition with New Essays
By Herman E. Daly

Sustainable Harvest and Marketing of Rain Forest Products
Conservation International

Taking Out the Trash: A No-Nonsense Guide to Recycling
By Jennifer Carless

Time for Change: A New Approach to Environment and Development
U. S. Citizens Network on UNCED

Trees, Why Do You Wait?
By Richard Critchfield

Turning the Tide: Saving the Chesapeake Bay
By Tom Horton and William M. Eichbaum

War on Waste: Can America Win Its Battle With Garbage?
By Louis Blumberg and Robert Gottlieb

Wetland Creation and Restoration: The Status of the Science
Edited by Mary E. Kentula and Jon A. Kusler

For a complete catalog of Island Press publications, please write:
Island Press, Box 7, Covelo, CA 95428, or call: 1-800-828-1302

Island Press Board of Directors